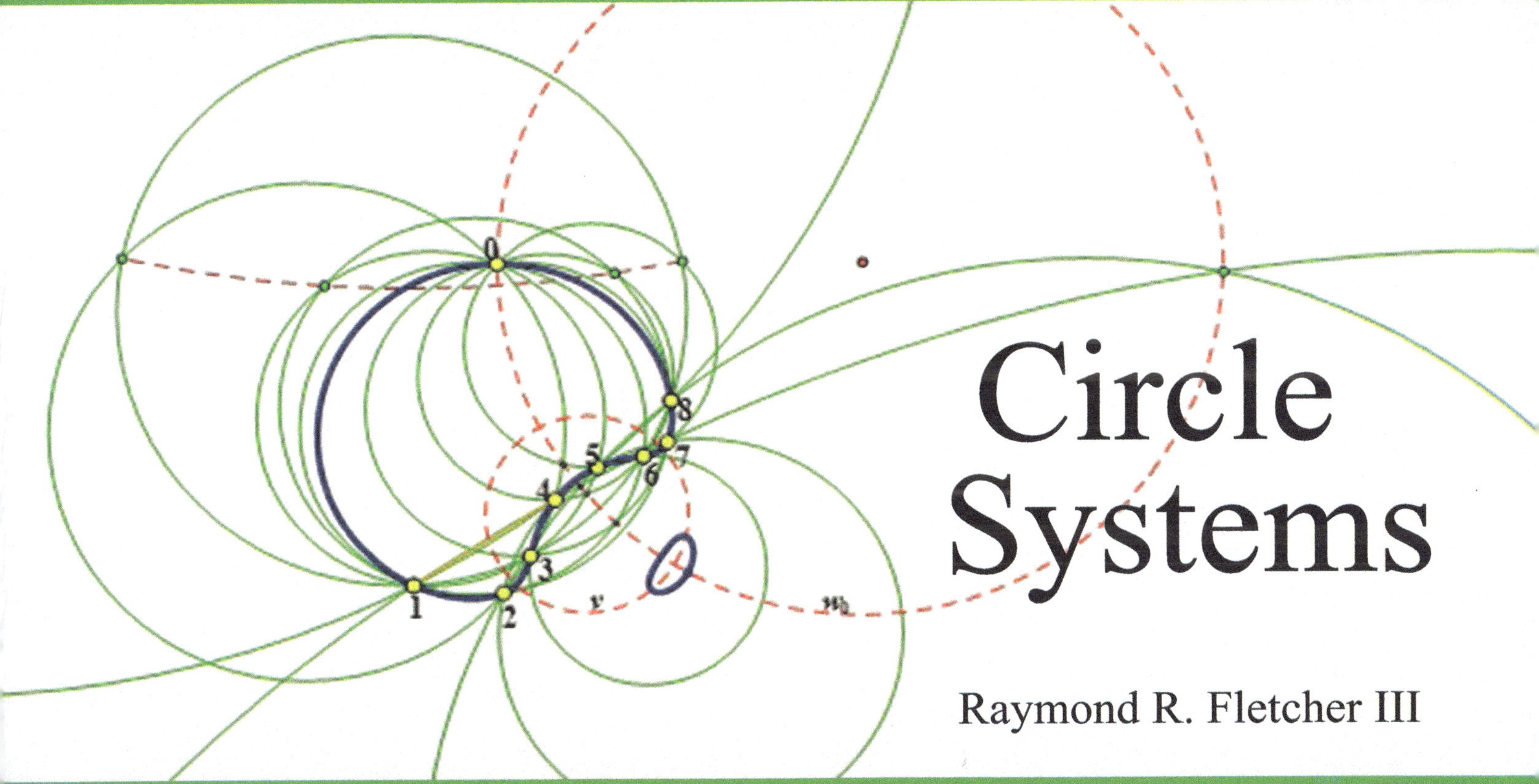

Circle Systems

Raymond R. Fletcher III

CIRCLE SYSTEMS

Raymond R. Fletcher III

Printed in the United States of America.

Library of Congress Control Number: 2022930707

ISBN	Paperback	978-1-68536-287-4
	eBook	978-1-68536-288-1

Westwood Books Publishing LLC
Atlanta Financial Center
3343 Peachtree Rd NE Ste 145-725
Atlanta, GA 30326

www.westwoodbookspublishing.com

Contents

Preface

A *circle system* in its original conception is primarily a combinatorial entity. It did not arise by reference to any previously studied object, but simply as a natural outcome of geometric experimentation assisted ably by the software *Geometer's Sketchpad.* The details of this evolutionary process involved the intermediary conceptions which we refer to as *perfect polygons* and *geometric triple systems.* These items are introduced and defined in the Introduction. Here we will define only our main subject. Let G be an abelian group and let g be a fixed element of G. Assign each element of G to points in the real projective plane so that for each four element subset {a,b,c,d} of G with a+b+c+d = g, the corresponding points are cocyclic. We refer to this structure as a *(G,g) circle system.* The abelian group G is the *base* and the element g is the *sum* of the system.

In Chapter 1 we use the properties of inversion to facilitate the construction and the determination of structural characteristics of two circle systems with base Z_8. The question of *equivalence* is introduced in connection with circle systems with base Z_n. This concept allows us to reduce our study of circle systems to just a few equivalence class representatives. A general construction of circle systems with base Z_n is obtained by a modification of a construction of a (Z,0) circle system given in [7].

In Chapter 2, the theory of geometric triple systems given in [2] is used to show that the vertices of any (Z,0) or (Z_n,0) circle system must lie on the inverse of an irreducible cubic curve which we call the *envelope* of the system. It is further shown that for these circle systems and for n sufficiently large, the cubic envelope inverts onto itself through two orthogonal circles. The theory of algebraic curves, in particular self-inversive cubic curves, thus adds a new dimension to our main object of study. The theory is further enriched by the existence of two related algebras which can be defined on the points of an irreducible cubic curve. If p, q are two points on an irreducible cubic curve C, then p*q denotes the third point besides p, q which lies on the line [p,q] and on C. This operation is used in [3] to define an abelian group on the points of C and to introduce the subject of elliptic curves. We define an analogous ternary operation δ on the points of a self-inversive cubic curve C by setting δ(a,b,c) equal to the fourth point besides a,b,c on C and on the circle (a,b,c). Any circle system with cubic envelope C is a subalgebra of C under this ternary operation. A simple formula is then provided which relates the two operations (*) and (δ). We show that certain geometric triple systems give rise to circle systems.

In Chapter 3 circle systems with base Z are constructed. Equivalence classes are determined and extensive use of properties of circle configurations proved in [6] are used for the constructions. The long inductions involved can be omitted on a first reading. In Chapter 4 the above mentioned algebraic operation (*) is used to construct a (Z_n,0) circle system on a given self-inversive cubic curve. Also the connection between perfect polygons and circle systems is explored.

In Chapter 5 the second major class of circle systems is introduced, namely those with base $Z_2 \times Z_{2m}$. We take the same approach to this subject as in the analysis of circle systems with base Z_n. We construct and prove the various circle systems with base $Z_2 \times Z_4$. Some more useful properties of inversion are developed to assist in the proofs. At this point no less than 8 circle systems with 8 points have been constructed. The problem of nonequivalence between these constructions is dealt with by considering various distinguishing parameters. These include *size, degree sequence; pair degree sequence, etc.,* all of which are defined in the Chapter.

In Chapter 6 general circle systems with base $Z_2 \times Z_{2m}$ are considered. Equivalence classes are determined. To determine nonequivalence, formulas for size (number of circle sets) and pair degree sequences are given. The proofs of these formulas is a significant aside and is reserved for Chapter 10. The most interesting part of this Chapter has to do with showing that certain circle systems with base $Z_2 \times Z_{2m}$ cannot exist with a cubic envelope due to the limitations imposed by the algebraic structure of the curve. At the end of Chapter 6 we examine the last known type of circle system, namely those with base $Z_2 \times Z_2 \times Z_2$.

We have already remarked that the points of a circle system Ω with a self-inversive cubic envelope C form a subalgebra (Ω,δ) of C under the ternary operation δ introduced in Chapter 2. In Chapter 7 a detailed investigation of the subalgebra structure of (Ω,δ) is given in case Ω has base Z_n. To facilitate this study a function $\varphi: C \rightarrow C$ defined by $\varphi(x) = \delta(x,x,x)$ is introduced and the structural properties of the corresponding digraph with arcs $\{x \rightarrow \varphi(x)\}$ are determined.

In Chapter 8 we consider the special case of circle systems all of whose points lie on a noncircular conic. A (Z,0) circle system is constructed on a noncircular conic and by a modification of this construction we obtain circle systems with base Z_n on any ellipse. It is also shown that the ternary operator δ satisfies the same identities on a noncircular conic as it does on a self-inversive cubic. In Chapter 9 we expand upon a Conjecture given in [6] regarding a concurrence of center lines for a *closed circle chain* consisting of 6 circles. Some special cases of this Conjecture are proved with the help of the δ operation. Except for references to the δ operation, Chapters 8 and 9 can be reviewed without examination of previous Chapters,

Chapter 10 focuses on counting arguments to determine the size and pair-degree sequence of circle systems with base Z_n and $Z_2 \times Z_n$. These formulas are useful in showing nonequivalence. Chapter 11 is a late addition which shows that one and two generator subalgebras of a self-inversive cubic curve with respect to the δ operation, also constitute circle systems.

Circle Systems is essentially a research monograph which represents the last instalment in a series of related topics by the Author including Perfect Polygons, [1] and Geometric Triple Systems [2]. To our thinking the main attraction of this work lies in the use of algebraic and graph theoretic methods for the study of geometric objects. It should have some appeal to mathematics or students of mathematics with a similar aesthetic sense. Besides Conjectures 28 and 52 in Chapters 3, 6 respectively, there are some interesting Conjectures given in Chapter 9 regarding the concurrence of center lines in various circle chains. These should offer adequate fodder for readers interested in further research. A further possibility for research involves the natural extension to the concept of a *conic system.* Let G be an abelian group and g a fixed element of G. Assign the elements of G to points in the projective plane such that for every 6 element subset $\{a,b,c,d,e,f\}$ with $a+b+c+d+e+f = g$, the corresponding points lie together on a conic.

Chapters 1, 2, 5, 8, 9 are sufficiently elementary for a College Senior in Mathematics to grapple with, but for most of the material graduate standing along with an interest in the application of algebraic tools to geometric structures is a likely prerequisite. Certainly experience in algebraic settings as well as some contact with inversive and projective geometry is helpful. Although attention to detail makes some of the proofs long and cumbersome, we consider it to be a distinguishing feature of the manuscript helpful for complete comprehension of the concepts used and the techniques applied.

I owe a debt of gratitude to the many teachers of mathematics I have had throughout my long career. I will mention here only my Masters Thesis Advisor Donald Silberger in whose tenure I experienced my first successful creation of original mathematics, and my PhD advisor Trevor Evans whose interest in algebras associated with combinatorics provided the essential background and flavor for my future work.

Ray Fletcher

Sutherlin Oregon

January 2019

Chapter 1

Introduction and Basic Properties

1.1 Introduction

This work represents the culmination of a series of studies which began with the concept of a *perfect polygon.* Let the elements of Z_n, (the integers mod n), be assigned to points in the projective plane and for each k ∈ Z_n let W_k denote the set of lines {[a, b] : a+b = k}. If for each k the lines in W_k concur at a point X_k then we call the set of points a *perfect n-gon P.* The set of points $\{X_k : k \in Z_n\}$ is called the *perspective set* of P. In Figure 82, (see Appendix), is illustrated a perfect pentagon. A perfect n-gon can be regarded as a projective generalization of a regular n-gon R: the vertices of R are labeled consecutively with elements from Z_n. We then find that the diagonals of R form n classes of parallel lines, each line in the same class corresponding to two points with the same sum. In this case the perspective points lie together on the line at infinity in the projective plane. Properties of perfect polygons are given in [1], where it is shown that either the vertices lie on a conic and the perspective points on a line, or the full set of vertices and perspective points lie on an irreducible cubic curve as in Figure 82. We call the cubic curve the *envelope* of the perfect polygon.

The next significant discovery was that if a perfect polygon P was constructed with three collinear perspective points then the entire perspective set of P formed collinear triples. The resulting figure is called a *closed perfect polygon.* (Figure 82 actually illustrates a closed perfect pentagon.) If ξ denotes the cubic envelope of P, then a binary operation (*) can be defined on the points of ξ by setting a*b equal to the unique point besides a, b which lies on the line [a,b] and on ξ. In this context a closed perfect polygon on ξ forms a subalgebra of (ξ,*) consisting of collinear triples of the form [a, b, a*b].

It was then observed that the points of P could be relabeled to form the next object in our series of studies: the *geometric triple system.* Let G be an abelian group and let g be a fixed element of G. Assign the elements of G to points in the projective plane. If for every three element subset {a,b,c} of G with a+b+c = g, the corresponding points are collinear, then we call these points and the corresponding triples a *(G,g) geometric triple system.* In [12] the necessary topics in modern abstract algebra including abelian groups, is adequately covered, and [3] is a good introduction to the projective plane. The $(Z_{10},0)$ geometric triple system corresponding to the closed perfect pentagon of Figure 82, is illustrated in Figure 83, (see Appendix). Properties of geometric triple systems are studied in [2] where it is shown that the points of a geometric triple system must also lie on a cubic curve. A natural extension of this definition determines our present subject. Let G be an abelian group and g ∈ G. Let $\phi: G \rightarrow \Pi$ be an injective mapping from G into the real projective plane Π such that no five points in φ(G) are cocyclic. If for each four element subset {a,b,c,d} of G with a+b+c+d = g, the corresponding points φ(a), φ(b), φ(c), φ(d) are cocyclic, then we call the set of points $\phi(G)$ and the associated cocyclic quadruples, a *(G,g) circle system*. The abelian group G is the *base* of the circle system; g is the *sum*, and the elements of G form the *point set* or *vertex set* of the system. The quadruples {a,b,c,d} with sum g are called *circle sets*. We will construct circle systems with various bases and sums, and determine their structural properties.

The theory of cubic curves will come again into play; this time we will find that circle systems have an envelope which is the inverse of a certain type of cubic. If ω is a circle with radius r and center O and P is a point different from O, then the *inverse* P' of P is the point on line OP such that $(OP)(OP') = r^2$. If we define a mapping ϕ on the plane minus the point O, by $\phi(P) = P'$, then ϕ is *involutory*, i,e., $\phi^2(P) = P$; ϕ fixes each point on ω; if P lies in the interior of ω, then P' lies in the exterior, and if P lies in the exterior of ω, then P' lies in the interior. The properties of inversion which we shall find useful are listed below. Proofs of these can be found in [9], [10], [11]. Two intersecting circles are *orthogonal* in case the tangents at the contact points are perpendicular. If a circle α is orthogonal to ω, then ϕ inverts the circle onto itself.

1.2 The $(Z_8,0)$ circle system

In Figure 1 we have illustrated a $(Z_8,0)$ circle system. The vertices are labeled with the corresponding elements from the base group. There are nine circles in this system labeled c_1, …, c_9 which correspond to the nine 4-element subsets of Z_8 with 0 sum. We say that the system has *size 9.* These circles and their corresponding circle sets are listed in the table below.

c_1:	1, 2, 6, 7
c_2 :	0, 3, 6, 7
c_3 :	0, 1, 2, 5
c_4 :	1, 4, 5, 6
c_5 :	2, 3, 5, 6
c_6 :	1, 3, 5, 7
c_7 :	0, 4, 5, 7
c_8 :	2, 3, 4, 7
c_9 :	0, 1, 3, 4

We have also illustrated two orthogonal circles w_0 and w_4 which do not belong to the system but are critical for its construction. The circle w_0 inverts all pairs of points {a,b} with a+b = 0, and w_4 inverts all pairs {c,d} with c+d = 4. In fact w_0 contains the two points 0, 4 which satisfy s+s = 0, and w_4 contains the two points 2, 6 which satisfy s+s = 4. The construction is accomplished as follows, (i) Draw orthogonal circles w_0, w_4; (ii) put points 7,5 as inverses with respect to w_4; (iii) put points 7, 1 as inverses with respect to w_0 ; (iv) let point 3 be the inverse of point 1 with respect to w_4; (v) put point 0 on w_0 and let point 4 be the inverse of point 0 with respect to w_4 ; (vi) draw circle (1,4,5) and let point 6 be one of the contact points of circle (1,4,5) with w_4; (vii) let point 2 be the inverse of point 6 with respect to w_0. To prove this construction we need the following properties regarding orthogonal circles and inversion can be found in [8]. Let ω be a circle with center O.

P1: If p, q are inverses with respect to ω, then any circle through p and q is orthogonal to ω.

P2: If p,q and s,t are two pairs of inverse points with respect to ω, then {p,q,s,t} is a cocyclic set of points.

P3: The inverse of a circle α not passing through O is a circle β not passing through O.

P4: The inverse of a circle α passing through O is a line not passing through O and perpendicular to the diameter of α through O. Conversely, the inverse of a line not passing through O is a circle through O.

P5: A circle α inverts with onto itself through ω iff α is orthogonal to ω.

P6: Inversion preserves angles.

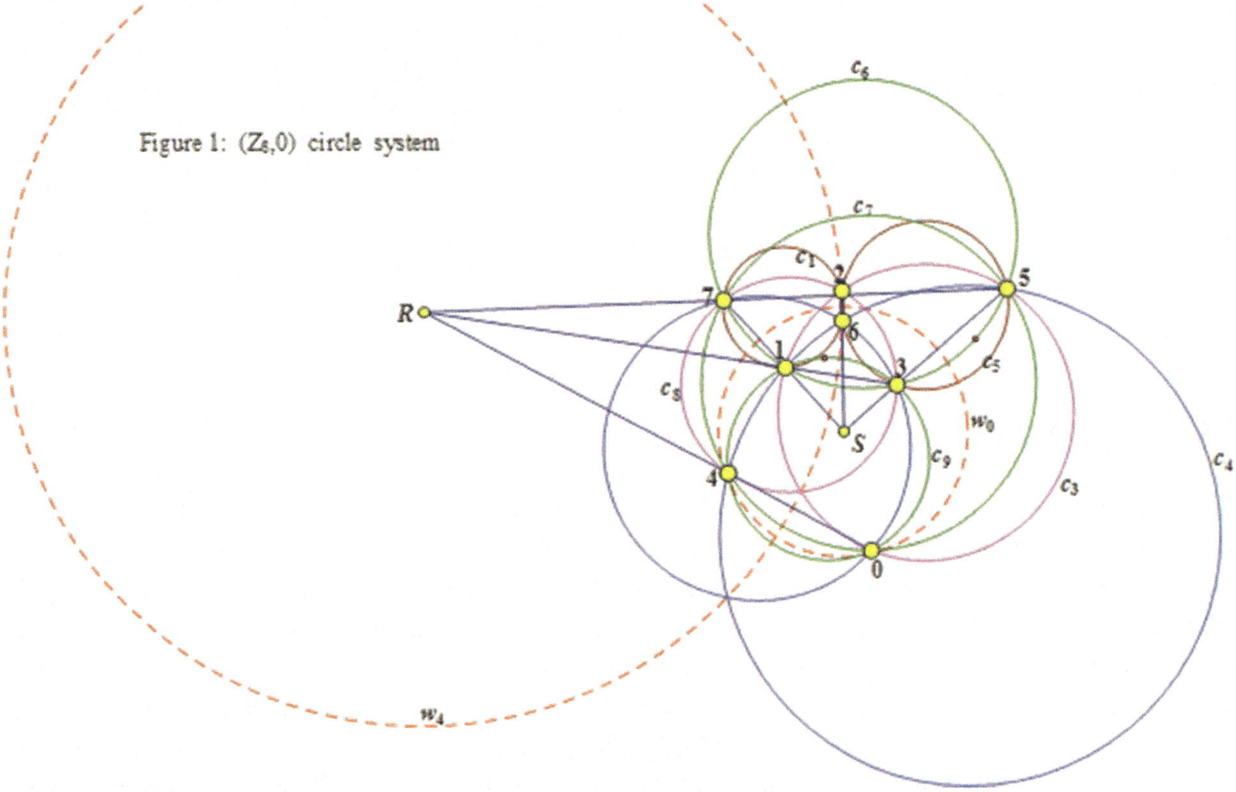

Figure 1: (Z_8,0) circle system

Also of use is the following result.

Lemma 1 *Let c_1, c_2 be orthogonal circles with centers J, K respectively. Let $\phi_1(x)$ denote the inverse of a point x with respect to c_1, and let $\phi_2(x)$ denote the inverse of a point x with respect to c_2. Then $\phi_1\phi_2 = \phi_2\phi_1$.*

Proof By P2 points x, $\phi_1(x)$, $\phi_2(x)$, $\phi_2\phi_1(x)$ lie together on a circle c_3 as in Figure 2. By P1, c_3 is orthogonal to both c_1 and c_2. By P4 the inverse of line [x, $\phi_1(x)$] is a circle c_4 containing K, $\phi_2(x)$ and $\phi_2\phi_1(x)$. Since [x, $\phi_1(x)$] meets c_1 at right angles, c_4 is orthogonal to c_1 by P6. By P5, $\phi_1(c_4) = c_4$ and $\phi_1(c_3) = c_3$. It follows that the point $\phi_1\phi_2(x)$ must lie on both c3 and c4. Since c3 ∩ c4 = {$\phi_2(x)$, $\phi_2\phi_1(x)$}, we can conclude that $\phi_1\phi_2(x) = \phi_2\phi_1(x)$. □

Now to the proof of our construction of a (Z_8,0) circle system. Circles c_6, c_7, c_9 are obtained by virtue of P2, and c4 is obtained by construction. By Lemma 1 points 3, 5 are inverses with respect to w_0. Since 2, 6 and 3, 5 are inverses with respect to w_0, we obtain c_5 by P2. By P3 we obtain circle c_2 as the inverse of c_4. Since point 6 lies on w_4, and w_0 is orthogonal to w_4 we must have, by P5, that point 2 lies on w_4. From this we obtain circle c_1 as the inverse of c_5 with respect to w_4. Circle c_3 is the inverse of c_2 with respect to w_0 and c_8 is obtained as the inverse of c_3 with respect to w_4. It should be mentioned that in a circle system we allow four collinear points to constitute a circle set; such a set of points can be regarded as belonging to a *circle* with infinite radius. We call this construction *perfect* since it can be constructed perfectly by straightedge and compass.

Figure 2: commutativity of inversion

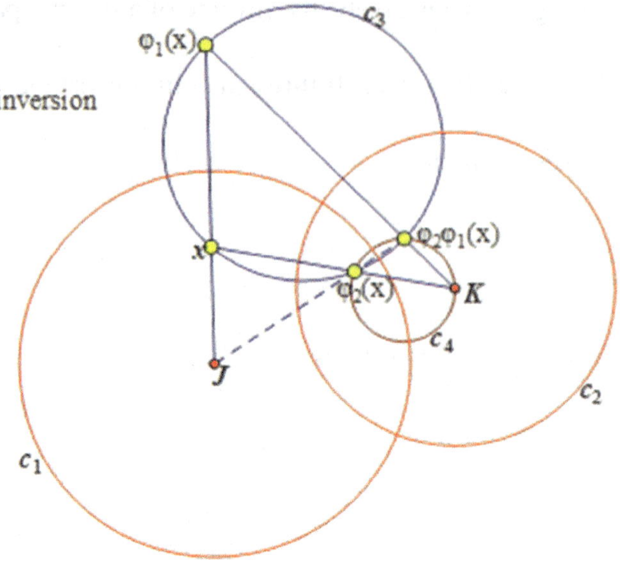

1.3 Circle Complexes

Our construction of the $(Z_8,0)$ circle system exhibits some important properties pertaining to a large variety of circle systems. Suppose Ω is a (G,g) circle system and there is an element s in G such that s+s = g. Let Q_s denote the set of pairs $\{\{p,q\}: p,q \in G, p+q = s\}$, then whenever $\{p_1, q_1\}$, $\{p_2, q_2\}$ are distinct pairs in Q_s, the set $\{p_1, q_1, p_2, q_2\}$ is a circle set of Ω. The collection of all such circle sets we denote by Ω_s and call the *circle complex with pair sum s.* We refer to the circle sets of Ω_s as *balanced*, and to any circle set not consisting of two pairs with equal sums, as *imbalanced.* The structural importance of a circle complex in Ω can be seen by considering three members: {a,b,c,d}, {a,b,e,f}, {c,d,e,f}, where a+b = c+d = e+f = s and s+s = g. There are two possible arrangements of the corresponding circles which we illustrate in Figure 3. In the first arrangement the radical center R of the three circles lies in the exterior of each circle. We refer to this formation as an *extroverted circle complex.* In the second arrangement the radical center R of the three circles lies in the interior of each circle. This formation we refer to as an *introverted circle complex.* In the extroverted complex there is a circle w with center R which is orthogonal to each of the three circles in the complex. Thus by P5, each circle inverts onto itself through w.

We define the *antiinverse* of a point P with respect to a circle w with center O, as the point obtained by reflecting the inverse of P with respect to w through the center O of w. For the introverted complex there exists a circle w with center R which antiinverts each of the three circles in the complex onto themselves. Suppose {h,k} is a pair of points in G different from {a,b}, {c,d} and {e,f} whose components also sum to s. The radical center of the circles corresponding to the circle sets {a,b,c,d}, {a,b.h,k} and {c,d,h,k} must also be R, the radical center of our original three circles. We can then say that the set of lines {[x,y] : x+y = s} all concur at a point R which we call the *radical center of the entire circle complex,* Ω_s . The radical center of Ω_s either occurs in the exterior of every circle in the complex, or in the interior of every circle in the complex. In the first case we call Ω_s *extroverted,* and in the second case we call Ω_s *introverted.* In case Ω_s is extroverted, there is a circle w_s with center R which is orthogonal to each circle in the complex and for which each pair of points {a,b} with a+b = s is an inverse pair. In case Ω_s is introverted, there is a circle

w_s with center R which antiinverts each circle in the complex onto itself and for which each pair of points {a,b} with a+b = s, is an antiinverse pair. In some cases Ω_s must be extroverted.

Figure 3: circle complexes

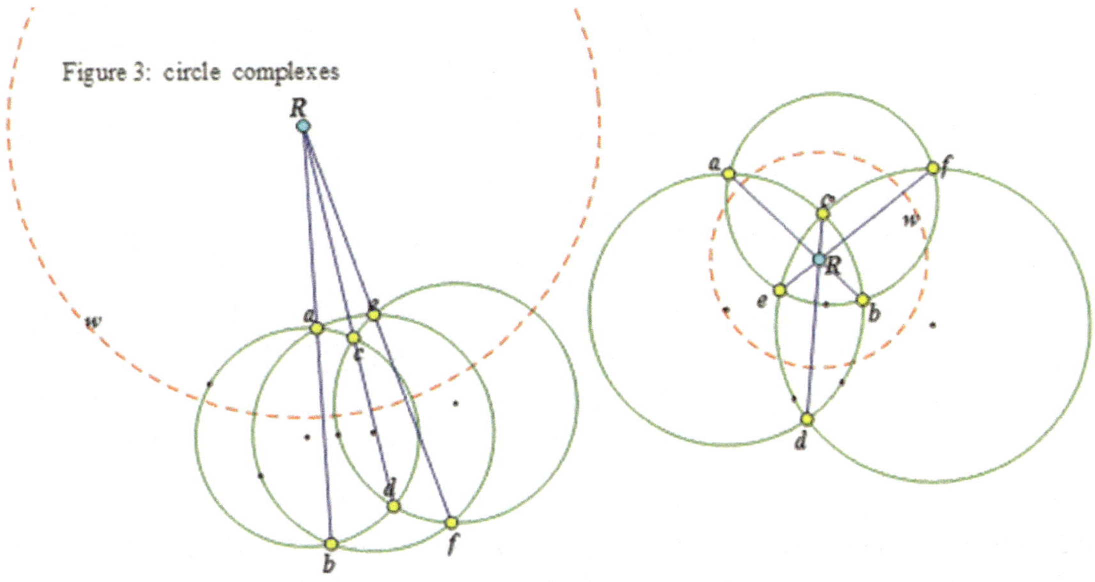

Theorem 2 *Let Ω be a (G,g) circle system and suppose there exists s ∈ G such that s+s = g. Also suppose* Q_s *contains three pairs {a,b}, {c,d}, {e,f} such that a+c+e = b+d+f and such that q = g – (a+c+e) ∉ {a,b,c,d,e,f}. Then* Ω_s *is extroverted and there exists a circle* w_s *which inverts each pair of points {p,q} with p+q = s.*

Proof Suppose Ω_s is introverted so that circles (a,b,e,f), (a,b,c,d), (c,d,e,f) contain their radical center in the interior of each circle as in Figure 4. The conditions a+c+e = b+d+f and q = g – (a+c+e) imply that (a,c,e,q) and (b,d,f,q) are circles of Ω. But no matter how the elements a,b,c,d,e,f are assigned to the six contact points of the three circles, we find that circles (a,c,e), (b,d,f) are disjoint. One such configuration is illustrated in Figure 4. It is thus impossible to construct the circles (a,c,e,q), (b,d,f,q) and we conclude that Ω_s is extroverted. □

As an application of Theorem 2, consider the (Z_8,0) circle system illustrated in Figure 1. We have g = 0 and there are two solutions to s+s = g, namely s = 0, 4. We thus obtain two circle complexes, Ω_0 and Ω_4. If we take a = 1; b = 7; c = 5; d = 3; e = 2; f = 6, we obtain q = 0 so the conditions of Theorem 2 are satisfied and we can conclude that Ω_0 is extroverted. If we take a = 1; b = 3; c = 4; d = 0; e = 5; f = 7, we obtain q = 6, and again the conditions of Theorem 2 are satisfied and we conclude that Ω_4 must also be extroverted. We obtain the following result as a Corollary to Theorem 2.

Figure 4: disjoint circles in introverted complex

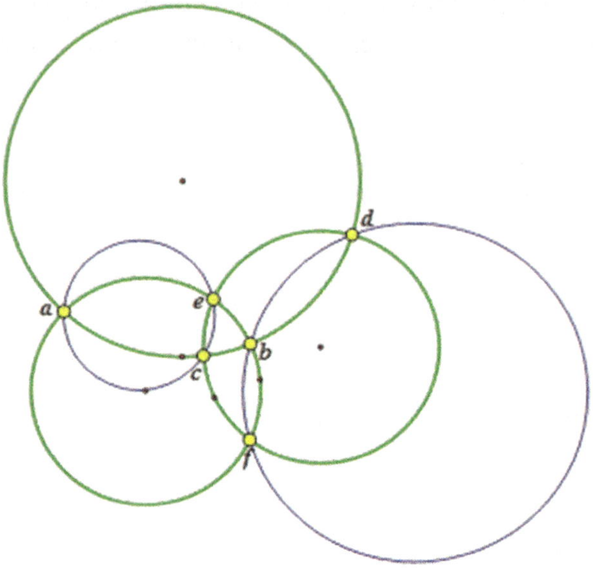

Theorem 3 *(i) If Ω is a (G,0) circle system and Q_0 contains distinct pairs {a,-a},{b,-b}, {c,-c} with a+b+c = 0, then Ω_0 is extroverted. (ii) If $n \geq 7$ and Ω is a $(Z_n,0)$ circle system, then Ω_0 is extroverted. (iii) If $m \geq 4$ is even and Ω is a $(Z_{2m},0)$ circle system, then Ω_m is extroverted.*

Proof (i) We have a+b+c = 0 = -a + -b + -c, and q = 0, so Ω_0 is extroverted by Theorem 2. (ii) Since $n \geq 7$, the elements 1, -1, 2, -2, 3,-3 of Z_n are distinct and we can take a = 1; b = 2 and c = -3 in (i). (iii) If m = 4, then Ω is a $(Z_8,0)$ circle system, and we have seen above that Ω_4 is extroverted. If $m \geq 6$ then the points $\left\{0,\ 1, \frac{m}{2}-1, \frac{m}{2}+1, m-1, m\right\}$ are distinct, and {0,m}, {1, m-1}, $\left\{\frac{m}{2}-1\ \frac{m}{2}+1\right\}$ are three pairs in Q_m. If Ω_m is introverted then circles $(1, m, \frac{m}{2}-1, \frac{m}{2})$, $(0, m-1, \frac{m}{2}+1)$ in Ω must be disjoint, but this is impossible since they contain the common point $\frac{m}{2}$. We conclude that Ω_m is extroverted. □

As in our construction of the $(Z_8,0)$ circle system, it is often the case that certain points must lie on the inverting circles.

Theorem 4 *Let Ω be a (G,g) circle system and suppose there exists s t∈ G such that s+s = g and t+t = s. If there exists distinct pairs {a, s-a}, {b, s-b}, {c, s-c} in Q_s such that a+b+c+t = g, then Ω_s is extroverted. Moreover, if w_s is the circle which inverts all pairs in Q_s then point t must lie on w_s.*

Proof If t = a, then a+a = s and {a, s-a} is not a pair. Similarly $t \neq$ b,c and thus α = (a,b,c,t) is a circle of Ω. If t = s-a, then s = t+t = (s-a)+(s-a) which implies s = 2a, again contradicting that {a, s-a} is a pair. Similarly $t \neq$ s-b, s-c and thus β = (s-a, s-b, s-c, t) is a circle in Ω. Now if Ω_s is introverted, then circles (a,b,c), (s-a, s-b, s-c) would be disjoint as in the proof of Theorem 2. Since these circles both contain point t, it must be that Ω_s is extroverted. Let w_s denote the circle which inverts all pairs in Q_s. Then circles α,β are an inverse pair with respect to w_s, and since point t lies on both circles, it must be that t lies on the inverting circle w_s. □

Corollary 5 *Let Ω be a $(Z_{2m},0)$ circle system with $m \geq 4$. Then Ω_0 is extroverted and there exists a circle w_0 which inverts all pairs of the form {t,-t}. Vertices 0, m must lie on w_0. There exists a circle w_m orthogonal to w_0 which inverts all pairs which sum to m. If m = 2q is even then Ω_m is also extroverted and Vertices q, q+m lie on w_m.*

Proof Ω_0 is extroverted by (ii) of Theorem 3, and Ω_m is extroverted by Theorem 3 (iii). So the inverting circles w_0 and w_m exist. By taking s = t = 0 in Theorem 4, we obtain that point 0 must lie on w_0, and by taking s = 0 and t = m in Theorem 4, we obtain that point m lies on w_0. By taking s = m and t = q in Theorem 4, we obtain that point q must lie on w_m, and by taking s = m and t = q+m in Theorem 4 we obtain that point q+m lies on w_m. Since 0, m are inverse points with respect to w_m and 0, m lie on w_0, it follows that w_0 is orthogonal to w_m by P1. □

With the completion of Corollary 5 we see that our construction of the $(Z_8,0)$ circle system is a general one.

1.4 Equivalence

The study of circle systems would be overwhelming if for each element g of an abelian group G an essentially different (G,g) circle system results. Luckily this is not the case due to the following concept of *equivalence.* A (G,g) circle system Ω is *equivalent* to an (H,h) circle system Γ in case there exists a bijection $\phi: G \rightarrow H$ such that whenever (a,b,c,d) is a circle of Ω, i.e., a+b+c+d = g, then $\phi(a) + \phi(b) + \phi(c) + \phi(d) = h$, i.e., $(\phi(a), \phi(b), \phi(c), \phi(d))$ is a circle of Γ. We call the mapping ϕ an *equivalence.* The concept of equivalence allows us to place circle systems in a small number of categories, but does not always differentiate between certain structural features. We also have found an example of two circle systems which are structurally identical but not equivalent. We determine next the equivalence classes of circle systems with base Z_n.

Theorem 6 *Let Ω be a circle system with base Z_n. (i) If gcd(n,4) = 1, then Ω is equivalent to a $(Z_n,0)$ circle system; (ii) If gcd(n,4) = 2, then Ω is equivalent to either a $(Z_n,0)$ circle system or a $(Z_n,1)$ circle system; (iii) If gcd(n,4) = 4 then Ω is equivalent to either a $(Z_n,0)$, $(Z_n,1)$ or $(Z_n,2)$ circle system. Moreover, the circle systems in (ii) and (iii) are mutually nonequivalent.*

Proof Let Ω be a $(Z_n,0)$ circle system, and for $k \in Z_n$ let Γ be a (Z_n,k) circle system. (i) If gcd(n,4) = 1, then 4 is a generator of Z_n, and thus the equation 4t = k has a solution for every $k \in Z_n$. Then the mapping $\phi:\Omega \rightarrow \Gamma$ given by $\phi(x) = x+t$ is an equivalence; (ii) If gcd(n,4) = 2 and $k \in Z_n$ then either $k \in \langle 2 \rangle$, (the subgroup of Z_n generated by 2), or $k \in 1+\langle 2 \rangle$. If $k \in \langle 2 \rangle$ then there exists $t \in Z_n$ such that 4t = k (mod n), and the mapping $\phi:\Omega \rightarrow \Gamma$ given by $\phi(x) = x+t$ is an equivalence. If $k \in 1+ \langle 2 \rangle$ then there exists $t \in Z_n$ such that 4t+1 = k (mod n), and the mapping $\phi:\Omega \rightarrow \Gamma$ given by $\phi(x) = x+t$ is again an equivalence; (iii) If gcd(n,4) = 4 and $k \in Z_n$ then k lies in one of the four cosets $\langle 4 \rangle$, $1+ \langle 4 \rangle$, $2+\langle 4 \rangle$, $3+\langle 4 \rangle$. If $k \in \langle 4 \rangle$ then there exists $t \in Z_n$ such that 4t = k (mod n); If $k \in 1+ \langle 4 \rangle$ then there exists $t \in Z_n$ such that 4t+1 = k (mod n); If $k \in 2+ \langle 4 \rangle$ then there exists $t \in Z_n$ such that 4t+2 = k (mod n), and if $k \in 3+ \langle 4 \rangle$ then there exists $t \in Z_n$ such that 4t+3 = k (mod n). In each case the mapping $\phi: \Omega \rightarrow \Gamma$ given by $\phi(x) = x+t$ is an equivalence. We have shown that every circle system with base Z_n is equivalent to either a $(Z_n,0)$, $(Z_n,1)$, $(Z_n,2)$ or $(Z_n,3)$ circle system.

The mapping $\psi:Z_n \rightarrow Z_n$ given by $\psi(x) = -x+1$ is an equivalence between a $(Z_n,3)$ and a $(Z_n,1)$ circle system, thus proving the first part of the Theorem. If n is even, say n = 2m, then a $(Z_n,1)$ circle system has no circle complexes since there is no element s in Z_n with s+s = 1. On the other hand, a $(Z_n,0)$ circle system Ω has circle complexes Ω_0 and Ω_m, and a $(Z_n,2)$ circle system has circle complexes Ω_1 and Ω_{m+1}. Thus a $(Z_n,1)$ system cannot be equivalent to a $(Z_n,0)$ or a $(Z_n,2)$ circle system. It remains to show, in the case gcd(n,4) = 4, that a $(Z_n,2)$ circle system Γ cannot be equivalent to a $(Z_n,0)$ circle system Ω . We do this by determining the size of the circle complexes in each system. If n = 4t then Ω has circle complexes Ω_0 and Ω_{2t}. Then Q_0 = {{1, 4t-1},{2, 4t-2}, …, 2t-1, 2t+1}}, and Q_{2t} = {{0,2t}, {1, 2t-1}, …, {t-1, t+1}} ∪ {{4t-1, 2t+1}, {4t-2, 2t+2}, …, 3t+1, 3t-1}}. Thus both Q_0, Q_{2t} have a total of 2t-1 pairs, and so Ω_0 and Ω_{2t} each consist of $\binom{2t-1}{2}$ circles.

The circle system Γ has circle complexes Γ_1 and Γ_{2t+1} with $Q_1 = \{\{1,0\}, \{2, 4t\text{-}1\}, \{3, 4t\text{-}2\}, \ldots, \{2t, 2t+1\}\}$ and $Q_{2t+1} = \{\{0, 2t+1\}, \{1, 2t\}, \{2, 2t\text{-}1\}, \ldots, t, t+1\}\} \cup \{\{4t\text{-}1, 2t+2\}, \{4t\text{-}2, 2t+3\}, \ldots,\{3t+1, 3t\}\}$, both with a total of 2t pairs. Thus Ω_0 and Ω_{2t} both have a total of $\binom{2t}{2}$ circles. Thus the circle complexes in Ω, Γ have different sizes with the consequence that a $(Z_n,2)$ circle system cannot be equivalent to a $(Z_n,0)$ circle system. □

1.5 The $(Z_8,2)$ circle system

In this Section we resume our construction of circle systems with base Z_8, in order to base our study on a wealth of simple examples. We will now present two constructions of a $(Z_8,2)$ circle system, one with two extroverted circle complexes, and the other with one complex introverted and the other extroverted. There are 10 circles in a $(Z_8,2)$ circle system which we list below.

c_1	(0,5,6,7)
c_2	(2,3,6,7)
c_3	(2,4,5,7)
c_4	(0,1,2,7)
c_5	(3,4,5,6)
c_6	(0,1,3,6)
c_7	(0,1,4,5)
c_8	(0,2,3,5)
c_9	(1,2,3,4)
c_{10}	(1,4,6,7).

First suppose Ω is a $(Z_8,2)$ circle system, whose circle complexes Ω_1 and Ω_5 are extroverted. Let w_1 denote the circle which inverts every pair {p,q} with p+q = 1, and let w_5 denote the circle which inverts every pair {s,t} with s+t = 5. Construct w_5 orthogonal to w_1 and put points 0,1 as inverses with respect to w_1. Let points 4, 5 be inverses of 1, 0 respectively with respect to w_5. By Lemma 1, points 4, 5 are inverses with respect to w_1. Let points 3, 6 be inverses with respect to w_1, and let 2, 7 be inverses of 3, 6 respectively with respect to w_5. By Lemma 1, points 2, 7 are inverses with respect to w_1. We then find that all ten circles of Ω are obtained by P1.

Now suppose that Ω_1 is introverted and let w_1 denote the circle which antiinverts every pair {p,q} with p+q = 1. Arrange the circles c_2, c_4 and c_6 so that their radical center R lies in the interior of each circle. Thus R is the center of w_1 as in Figure 5. Let point 5 denote the intersection of circles (0,6,7), (0,2,3) different from point 0, and let point 4 be the intersection point of circles (1,7,6), (3,5,6) different from 6. The circles c_1, c_2, c_4, c_5, c_6, c_8, c_{10} are obtained immediately by construction, so it remains only to prove circles c_3, c_7, c_9. To accomplish this we will make use of the following properties of antiinversion. Let w be a circle with center R.

P7: If {p,q} and {s,t} are antiinverse pairs with respect to w, then {p,q,s,t} is a cocyclic set.

P8: If α is a circle not passing through R, then the antiinverse of α with respect to w is a circle β not passing through R.

P9: If p,q are antiinverses with respect to w, then any circle through p,q is antiinverted onto itself with respect to w.

P10: antiiversion preserves angles.

By P9 circle (3,5,6) antiinverts onto itself, and thus the antiinverse of point 5 lies on (3,5,6). By P8 the antiinverse of circle c_8 = (0,2,3,5) is circle (1,7,6) and thus the antiinverse of point 5 must lie on (1,7,6). Consequently the antiinverse of 5 lies at the intersection of circles (3,5,6) and (1,7,6), i.e., point 4 is the antiinverse of point 5. Then {2,4,5,7} consists of two antiinverse pairs, and thus c_3=(2,4,5,7) is a circle by P7. Also by P7 we obtain circle c_7. Finally, c_9 is the antiinverse of c_1, so we obtain c_9 by P8. This perfect construction is illustrated in Figure 5. If we add 6 to each point in Figure 5, we obtain again a $(Z_8,2)$ circle system, but now Ω_5 is the introverted complex and Ω_1 is extroverted. It is interesting to note that w_5 antiinverts onto itself through w_1. This is so since antiinversion preserves angles (P10): since w_5 is orthogonal to each circle of Ω_5, the antiinverse of w_5 with respect to w_1 must be orthogonal to the antiinverse of each circle of Ω_5. But Ω_5 itself constitutes the full set of antiinverses of circles in Ω_5, so the antiinverse of w_5 must be orthogonal to every circle in Ω_5 and this forces the antiinverse of w_5 to be w_5 itself.

We also observe in Figure 5 the concurrence of the circles c_4, c_5, w_5 at the points labeled u, v and that line [u,v] passes through R, the center of w_1. This can be explained as follows. Circles c_4 and c_5 are inverses with respect to w_5, i.e., w_5 is a midcircle for c_4, c_5 and thus w_5 contains the contact points u, v. Let x denote the antiinverse of u with respect to w_1. Since u lies on c_5 and c_5 antiinverts onto itself via w_1, it must be that x lies on c_5, and since u lies on c_4 and c_4 antiinverts onto itself via w_1, it must be that x lies on c_4. Thus x is an intersection point of c_4, c_5 which forces x = v. Then u, v are antiinverses with respect to w_1 and thus segment [u,v] passes through the center R of w_1. If p, q denote the contact points of c_3, c_6 then we obtain similarly that w_5 contains p, q and that p,q are antiinverses with respect to w_1. It is not clear as yet how *isomorphism* should be defined in case of circle systems, but certainly our two constructions of $(Z_8,2)$ circle systems, though equivalent, should not be considered isomorphic. Some general results can be extracted from our study of $(Z_8,2)$ circle systems.

Figure 5: $(Z_8,2)$ circle system with antiinverter

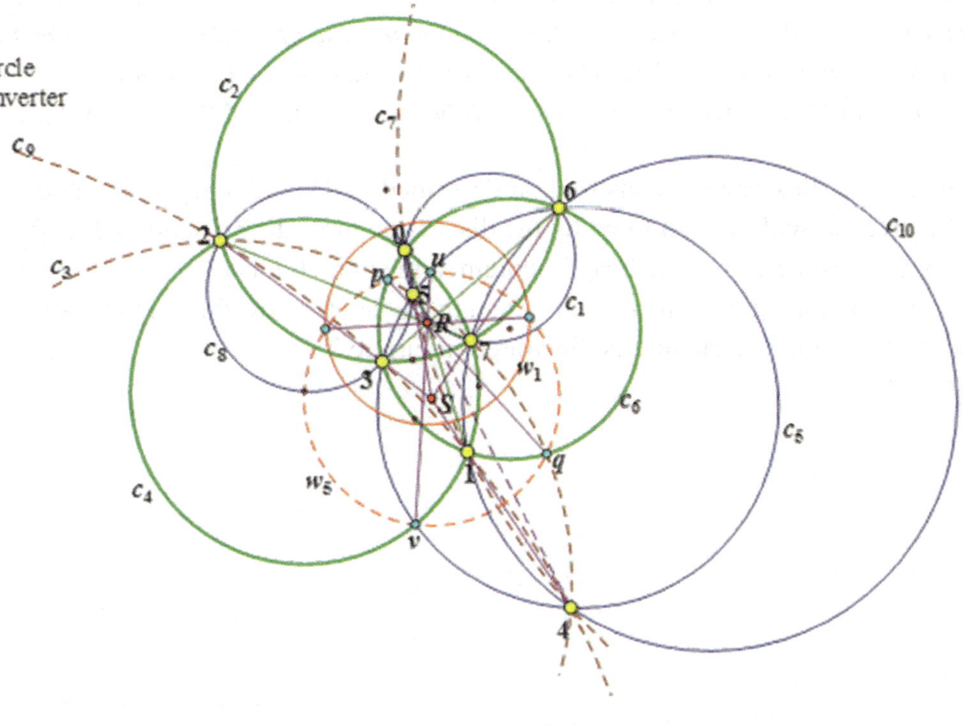

Theorem 7: *Let Ω be a (G,g) circle system and suppose there exist distinct elements s, t in G such that s+s = t+t = g, and suppose Ω_s is introverted. If there exists a $\in$ G such that elements a, t-a, s-a, s-t+a are distinct, then Ω_t must be extroverted. Moreover, if w_s is the circle which antiinverts every pair of points with sum s, and w_t is the circle which inverts every pair of points with sum t, then w_t antiinverts onto itself via w_s.*

Proof Since {a, s-a} and {t-a, s-t+a} are antiinverse pairs with respect to w_s,it must be that lines [a,s-a], [t-a, s-t+a] meet at the center R of w_s. Since (a, t-a, s-a, s-t+a) is a circle α of Ω which antiinverts onto itself via w_s, it must be that R lies in the interior of this circle. For this to happen the points of α must occur either in the order (a, t-a, s-a, s-t+a) or in the order (a, s-t+a, t-a, s-a) as α is traversed clockwise. In any case the lines [a, t-a], [s-a, s-t+a] meet at a point S exterior to α. Since the pairs {a, t-a}, {s-a, s-t+a} lie in $Q_{t,}$ the circle α also lies in the circle complex Ω_t. As the point S is exterior to α, Ω_t must be extroverted. Let ψ_s (x) denote the antiinverse of x with respect to w_s. If β is a circle in Ω_t then β is orthogonal to w_t, and by P10, ψ_s (β) is orthogonal to ψ_s (w_t) . If β = (a, t-a, b, t-b) then ψ_s (β) = (s-a, s-t+a, s-b, s-t+b) which consists of two pairs which sum to t. Thus ψ_s (β) $\in$ Ω_t and ψ_s (Ω_t) = Ω_t. Consequently ψ_s (w_t) is orthogonal to every circle in Ω_t, forcing ψ_s (w_t) = w_t. □

The construction and proof of a (Z_8,1) circle system will be given in Chapter 4.

1.6 Construction of (Z_n,0) circle systems

In this Section we progress from the concrete examples of circle systems with base Z_8 to the general construction of circle systems with base Z_n for any positive integer n. The method used is based on the construction of a circle system with base the full set Z of integers. In [7] the following construction of a (Z,0) circle system is proved.

Theorem 8 *Let w_0 be a circle and put vertex 0 on w_0. Put vertices 1, 2 in the plane and not on w_0 so that 1,2 are not inverses with respect to w_0, and let -1, -2 be their inverses respectively with respect to w_0. Put vertex 3 on circle (0,-1,-2) and not on w_0, and let vertex -3 be the inverse of 3 with respect to w_0. In general, if vertices {0, ± 1, ± 2, ..., ± k} have been determined, let vertex k+1 be the intersection point besides -k of circles (-k, -1,0), (-k, -2,1), and let vertex -k be the inverse of vertex k with respect to w_0. Let Ω denote the set of integer labeled points defined in this way. We assume the generators 1, 2, 3 of Ω are chosen so that no two points of Ω receive the same integer label. Then Ω* Is a (Z,0) circle system. □

In Figure 6 we illustrate the construction described in Theorem 8. Note that all pairs of the form {q,-q} must be inverses with respect to w_0 by the discussion preceding Theorem 2, and by Theorem 2, the circle complex Ω_0 must be extroverted. By Theorem 4, the point 0 must lie on w_0 as in our construction. We will be interested in the figure which results when the generators are shifted to make vertex n lie exactly on w_0. For n = 5 the resulting structure is illustrated in Figure 7.

Figure 6: (Z,0) circle system

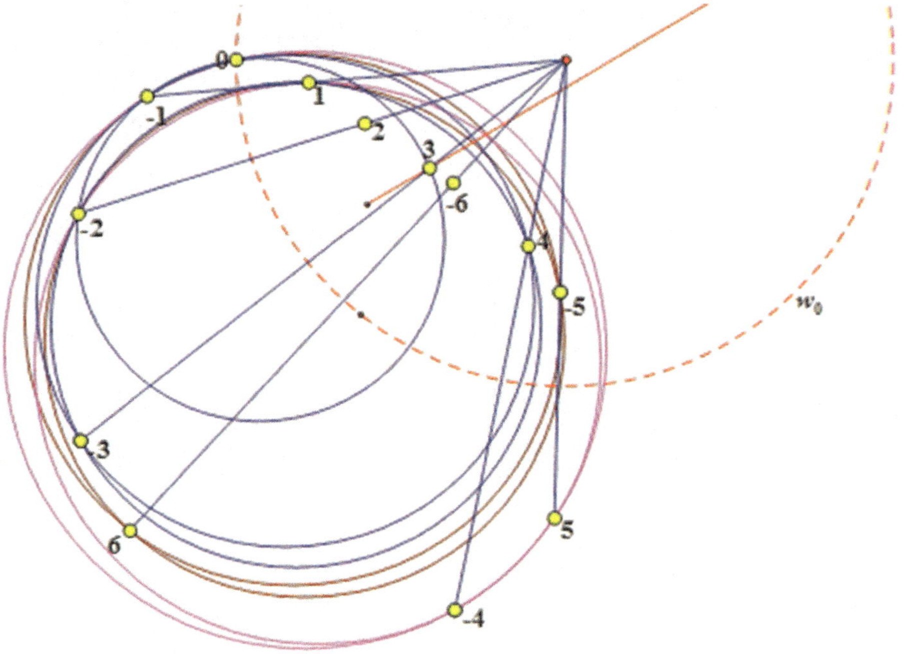

Theorem 9 *In the above construction of a (Z,0) circle system, when the generator(s) are shifted so that vertex m lies on w_0, then a $(Z_{2m},0)$ circle system results.*

Proof Since m lies on the circle of inversion and -m is constructed as the inverse of vertex m, it must be that m, -m represent the same point. Thus vertex m+1 which is by definition the intersection of the circles (-m,-1,0), (-m, -2,1) is at the same time the intersection of the circles (m,-1,0), (m, -2,1) which is vertex -m+1. It follows that the inverse of vertex m+1, i.e., -(m+1) is identical with the inverse of vertex -m+1, i.e., m-1. Assume inductively that for some positive integer t we have labels m+t, -m+t ascribed to the same point, and thus labels -m-t, m-t ascribed to the same point. Then vertex m+t+1 = (-m-t, -1,0)∩ (-m-t, -2,1) = (m-t,-1,0)∩ (m-t,-2,1) = -m +t+1. By induction we can conclude that for any positive integer k, the same point receives the labels m+k and -m+k. More generally if a, b ∈ Z and a ≡ b (mod 2m), then, in the coalesced (Z.0) system the same point receives both a, b as labels. If we let Ω denote the coalesced (Z,0) system then Ω has only the 2m distinct points belonging to Z_{2m}. To see that Ω is in fact a $(Z_{2m},0)$ circle system, consider that none of the (Z,0) system circles are actually lost, they are simply coalesced as vertex m is shifted towards its resting place on w_0. Thus a circle (a,b,c,d) in the (Z,0) system is identical with circle (a_1, b_1, c_1, d_1) whenever $a_1 \equiv a$, $b_1 \equiv b$, $c_1 \equiv c$ and $d_1 \equiv d \pmod{2m}$. As a+b+c+d = 0, we obtain $a_1 + b_1 + c_1 + d_1 = 0 \pmod{2m}$, and thus every circle in Ω is obtained. □

In Figure 7 a $(Z_{10},0)$ circle system, obtained in the way of Theorem 9, is illustrated.

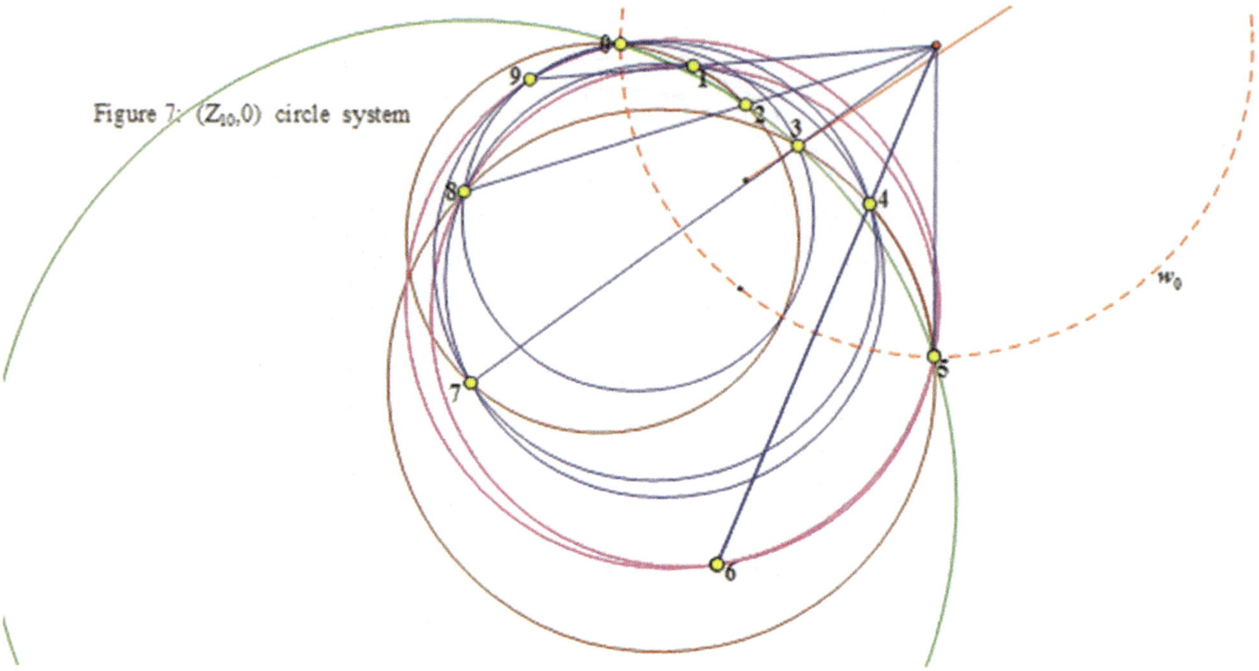

Figure 7: (Z_{10},0) circle system

Starting with the (Z,0) circle system, we can also shift vertex 3 so that points 4, -5 coincide. It can then be shown that the (Z,0) system coalesces to a (Z_9,0) circle system as illustrated in Figure 8. If two labels a, b are congruent mod 9, then they label the same point. While it is clear that when vertex 5 lies on w_0, we must have that point 5 and -5 coincide, it is not clear that it should be possible to make points 4, -5 coincide. This will be demonstrated in the next Chapter by showing that all the points of any (Z,0) system must lie on an irreducible algebraic curve.

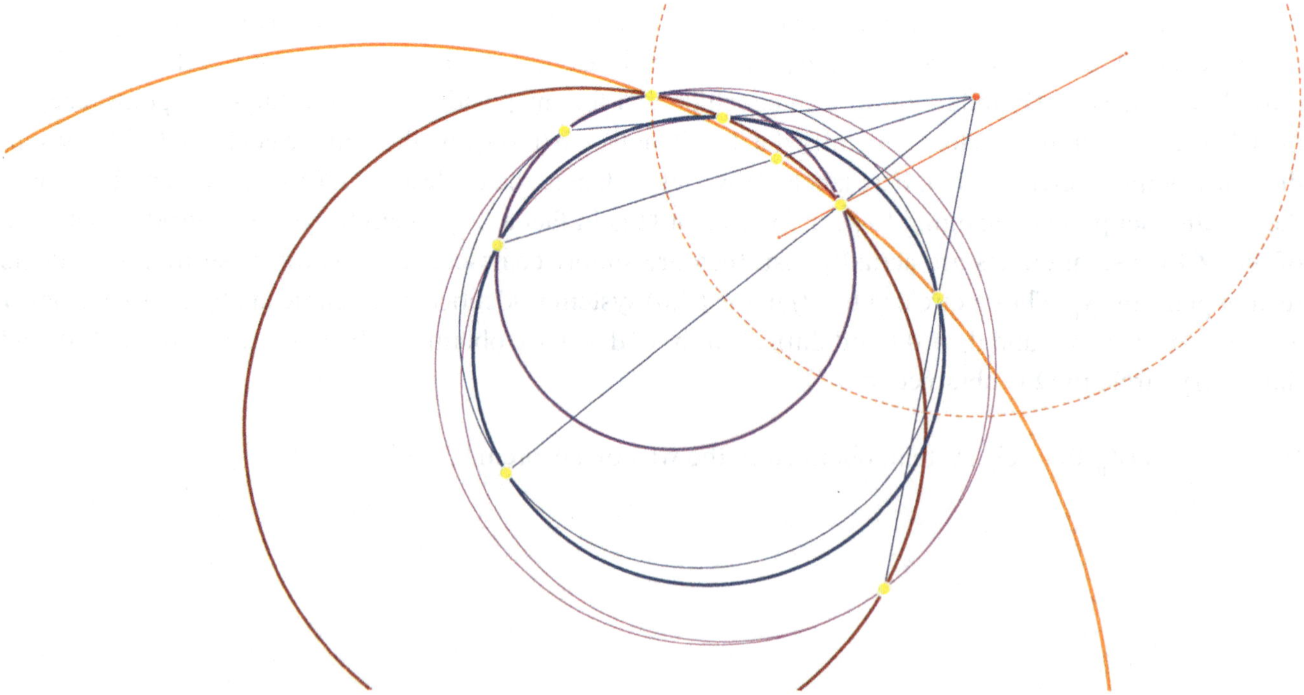

Chapter 2

The Envelope of a Circle System

2.1 Geometric triple systems and circle systems

At the very end of the previous Chapter it was hinted that the vertices of a circle system must lie on an algebraic curve and that this property would allow a more satisfactory construction. At this point the concept of a *geometric triple system,* which was briefly introduced in Section 1.1, comes into play. Here we will review the definition of a geometric triple system and show how this concept arises naturally in the study of circle systems.

Suppose Ω is a (Z,0) circle system and we invert by a circle v with center at vertex 0. By P4 every circle of Ω of the form (0,a,b,c) inverts into a line containing the triple of points {a′,b′,c′} where a+b+c = 0. Thus every triple of nonzero points a′,b′,c′ with a+b+c = 0 is collinear. There remains triples of the form {a′, -a′, 0′} where 0′ must be defined. To make this definition we will make use of the following additional property of inversion.

P11: Let α, β be two circles and let $\phi_\alpha(x)$ denote the inverse of a point x with respect to α. If {p,q} is an inverse pair with respect to β, then $\{\phi_\alpha(p), \phi_\alpha(q)\}$ is an inverse pair with respect to $\phi_\alpha(\beta)$. If β passes through the center of α, then $\phi_\alpha(\beta)$ is a line L which passes through the contact points of α, β and $\phi_\alpha(p)$, $\phi_\alpha(q)$ are reflections of each other across L.

Now since the circle w_0 which contains vertex 0, passes through the center, (vertex 0) of v, it follows by P11 that $\phi_v(a) = a'$ and $\phi_v(-a) = -a'$ are reflections of each other across the line $\phi_v(w_0) = w_0'$. Thus all lines of the form [a′, -a′] are perpendicular to w_0' hence parallel to each other. Then all such lines meet at the same point at infinity in the projective plane. We define this point at infinity to be 0′. With this definition the set of points $\phi_v(\Omega)$ becomes a *(Z,0) geometric triple system.* A (G,g) geometric triple system is defined in [2] as follows. Let G be an abelian group with $g \in G$, and let $\psi: G \to \Pi$ be an injective mapping such that no four points in $\psi(G)$ are collinear. If for every three element subset {a,b,c} of G with a+b+c = g the corresponding points $\psi(a)$, $\psi(b)$, $\psi(c)$ are collinear, then we call the set of points $\psi(G)$ along with the associated collinear triples a *(G,g) geometric triple system with base G and sum g.* The significant result given in [2] is the following. An algebraic curve F(x,y) = 0 is *irreducible* if no nontrivial factorization of F exists.

Theorem 10 *The vertices of any geometric triple system with base Z or Z_n lie on an irreducible cubic curve.* □

Similarly, if Ω is a $(Z_n,0)$ circle system and we invert by a circle v with center at vertex 0, we obtain a $(Z_n,0)$ triple system. In Figure 9 we have included the inverse of the $(Z_9,0)$ circle system with primed vertices.

The triples in the resulting $(Z_9,0)$ triple system are indicated with bold lines. Triples containing the point 0' at infinity are indicated by the set of mutually parallel lines. As a consequence of Theorem 10 we have:

Theorem 11 *The vertices of a (Z,0) or any $(Z_n,0)$ circle system lie on the inverse of an irreducible cubic curve.* □

The curve described in Theorem 11 is necessarily algebraic, and we shall refer to it as the *envelope* of the circle system. Theorem 11 can be used to explain the construction of a $(Z_n,0)$ circle system when n = 2m+1 is odd. As vertex 3 is shifted around the fixed circle (0,1,-2) of a (Z,0) circle system, we obtain a series of (Z,0) circle systems each with its own envelope, and in which vertices m and -m-1 proceed in opposite directions along the series of envelopes. Thus points m and -m-1 can be made to coincide and a $(Z_{2m+1},0)$ circle system results.

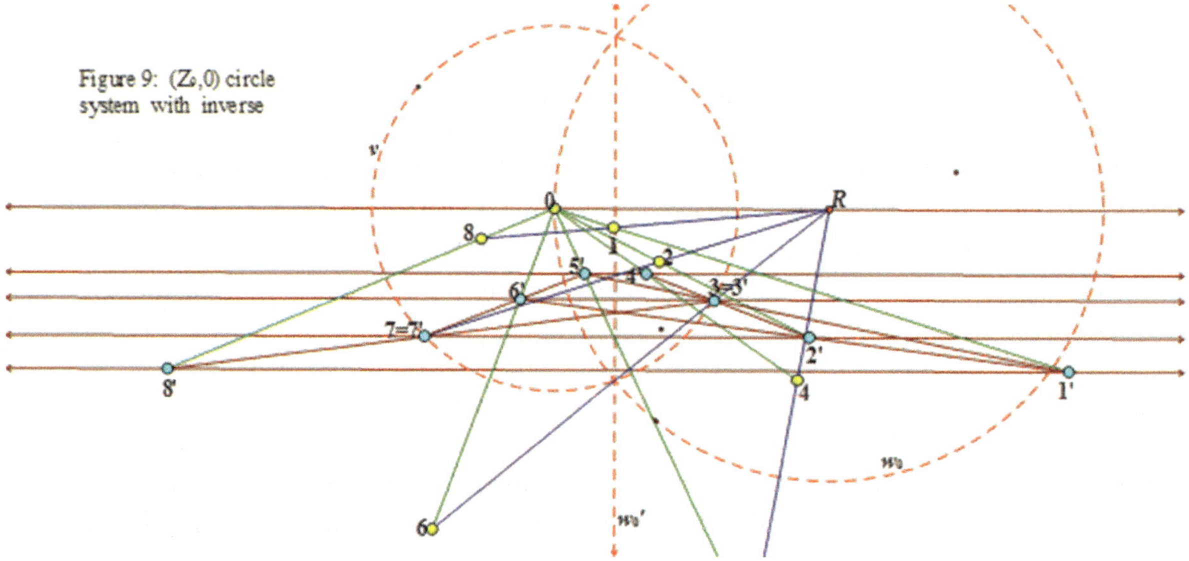

Figure 9: $(Z_9,0)$ circle system with inverse

We can actually say something more about the envelope of a (Z,0) or a $(Z_n,0)$ circle system. It is based on the following Lemma which tells about the nature of a cubic curve which contains so many reflective pairs.

Lemma 12 *Let γ be an irreducible cubic curve which contains at least five pairs of points which are reflections of each other across the y-axis. Then γ must be symmetric with respect to the y-axis.* □

Proof The general cubic γ has the equation:

$$(1) \quad f(x,y) = ax^3 + bx^2y + cxy^2 + dy^3 + ex^2 + fxy + gy^2 + hx + ky + j = 0.$$

If (s,t) and (-s,t) is a pair of points on γ, then we obtain

$$(2) \quad f(s,t) - f(-s,t) = 2as^3 + 2cst^2 + 2fst + 2hs = 0.$$

Four more pairs on γ which are reflections across the y-axis, yield four more equations like (2) in the four coefficient variables a, c, f, h. We then have five equations in four variables with the consequence that a = c = f = h = 0. Then the equation of γ reduces to

$$(3) \quad f(x,y) = bx^2y + dy^3 + ex^2 + gy^2 + ky + j = 0.$$

Now f(x,y) = f(-x,y) so γ is symmetric with respect to the y-axis. □

Theorem 13 *Let Ω be a (Z,0) or a (Z_n,0) circle system with $n \geq 11$ and let w_0 denote the circle which inverts all pairs of the form {q,-q} and which contains vertex 0. Then the envelope of Ω inverts onto itself via w_0. If n = 2m, then the envelope of a (Z_{2m},0) circle system inverts onto itself via the two orthogonal circles w_0 and w_m.*

Proof Let v be a circle with center at vertex 0 and invert Ω through v to obtain a (Z,0) or (Z_n,0) triple system Γ. Let γ denote the irreducible cubic curve which contains the points of Γ, in accordance with Theorem 10. Since $n \geq 11$, Γ will contain at least five pairs of points which are reflections across the line $\phi_v(w_0) = w_0'$, the inverse of w_0 with respect to v. We may take w_0' to be the y-axis and then apply Lemma 12 to conclude that γ is symmetric with respect to the line w_0'. It must then be, by P11, that the envelope of Ω, i.e., $\phi_v(\gamma)$ inverts onto itself via the circle $\phi_v(w_0') = w_0$. In case Ω is a (Z_{2m},0) circle system and w_m is the circle which inverts all pairs {p,q} with p+q = m, let u be a circle with center m and invert Ω via u. Every circle (a,b,c,m) of Ω which contains the point m will invert to a line [a,b,c] with a+b+c = m, and thus a (Z_{2m},m) triple system Γ is obtained. By Theorem 10 the points of Γ lie on an irreducible cubic curve γ. Since $n \geq 11$, Γ will contain at least five pairs of points which are reflections across the line $\phi_u(w_m) = w_m'$, the inverse of w_m with respect to u. It follows again by Lemma 12, that γ is symmetric with respect to the line w_m', and the envelope of Ω, i.e., $\phi_u(\gamma)$ inverts onto itself via w_m. So,in this case Ω inverts onto itself via both circles w_0, w_m which are shown to be orthogonal in Corollary 5. □

Theorem 13 is illustrated in Figure 10 with a (Z_{12},0) circle system. Note that $\phi_v(w_6)$ is a circle which inverts the cubic γ onto itself. We call γ *self-inversive*. Even if n is odd, we can show that there exist two orthogonal circles which invert the envelope of a (Z_n,0) circle system onto itself. To do this we make use of the following Theorem which is proved in [7]. A *singular point* of an algebraic curve F(x,y) = 0 is a point on the curve, e,g, a cusp, which has no unique tangent line. F is *nonsingular* in case it contains no singular points.

Figure 10: (Z_{12},0) circle system inverted to obtain a (Z_{12},0) triple system

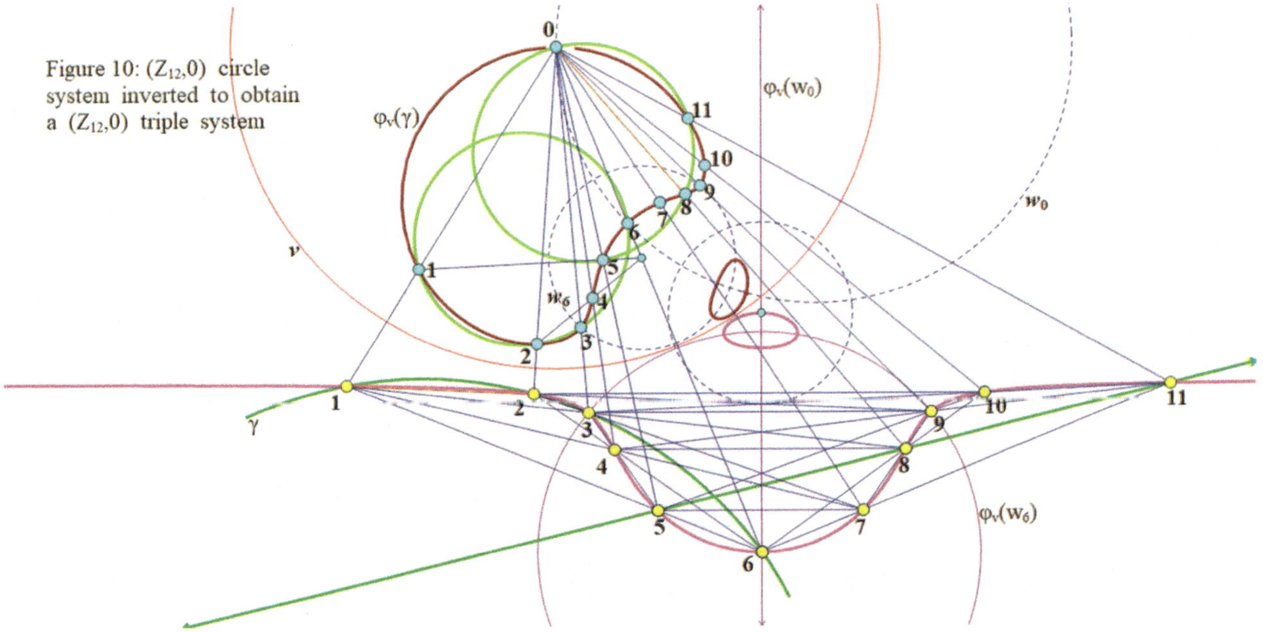

Theorem 14 *Let γ be a nonsingular irreducible cubic curve which inverts onto itself via a circle w. Then there exists a second circle v orthogonal to w which also inverts γ onto itself.* □

Theorem 14 is illustrated in Figure 11. The center X of w must lie on the cubic γ. In case w meets γ in two points p, q, we let Y denote the point besides p, q where line [p,q] meets γ. We then draw circle v with center Y orthogonal to w.

Theorem 15 *Let Ω be a $(Z_n,0)$ circle system with n odd and let w_0 denote the circle which inverts the envelope δ of Ω onto itself. Then there exists a circle v orthogonal to w_0 which also inverts δ onto itself.*

Proof Let u be a circle with center at vertex 1 of Ω, and $\gamma = \phi_u(\delta)$ denote the inverse of δ via u. All circles of Ω of the form (a,b,c,1) are inverted to lines containing the inverses a′, b′, c′ of a,b,c respectively where a+b+c = -1. Thus γ is the envelope of a $(Z_n,-1)$ triple system and must be an irreducible cubic curve by Theorem 10. Since w_0 inverts δ onto itself, the circle $\phi_u(w_0) = w_0'$ inverts the cubic γ onto itself . Now by Theorem 14 there exists a circle v′ orthogonal to w_0' which also inverts γ onto itself. Then circle $v = \phi_u(v')$ inverts δ onto itself and is orthogonal to w_0. □

Theorem 15 is illustrated in Figure 12 with a $(Z_9,0)$ circle system.

Figure 11: Orthogonal circles which invert a cubic curve onto itself

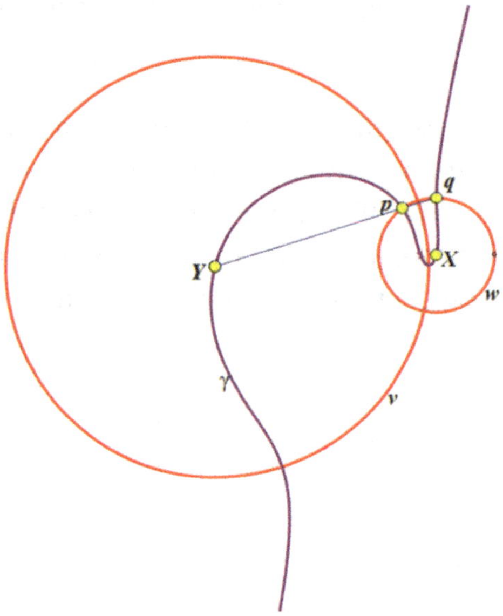

Figure 12: $(Z_9,0)$ circle system with envelope and inverting circles

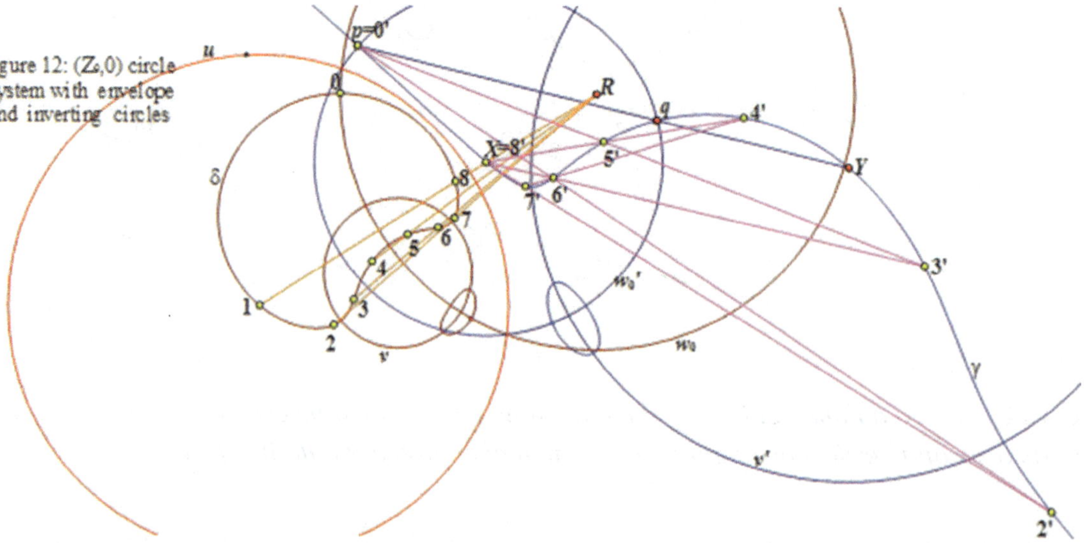

2.2 Using (G,0) triple systems to construct (G,0) circle systems

Some further properties of self-inversive cubic curves will be required for this Section. In [3] irreducible cubic curves are put into five categories. The first three categories, (Types I,II,III), cover cubic curves which contain singular points. We will confine our attention to the nonsingular cubic curves which comprise cubics of Type IV and V. Type IV consists of those cubics whose equations can be transformed into $y^2 = x(x-1)(x-w)$ where $w>1$, and Type V consists of those cubics whose equations can be transformed into $y^2 = x(x^2 + kx + 1)$, $-2<k<2$. Type V cubics consist of a single connected component as in Figure 11, and Type IV cubics consist of two connected components which we call the *bell* and the *oval.* If γ is a Type IV cubic then the bell is a subalgebra of $(\gamma,*)$ and the oval is *anticlosed,* i.e., if $a,b \in$ oval than $a*b \in$ bell. Figures 10, 13, 14 feature Type IV cubics. A Type V cubic which inverts onto itself, inverts onto itself through two mutually orthogonal circles, and a Type IV cubic which inverts onto itself, inverts onto itself through three mutually orthogonal circles, and antiinverts onto itself through a fourth circle. In this connection there are two related algebraic systems which we introduce. Let γ be an irreducible cubic curve and suppose p, q are two nonsingular points on γ. We can then define a binary operation (*) on γ by setting p*q equal to the third point besides p,q on line [p,q] and on γ. The expression p*p denotes the point besides p, on the tangent to γ at p and on γ. It is possible to have p*p = p, in which case p is called a *flex point* of γ. This binary operation satisfies the following identities.

(i) $p*q = q*p$
(ii) $p*(p*q) = q$
(iii) $(p*q)*(r*s) = (p*r)*(q*s)$

Identities (i), (ii) are clear but identity (iii) requires some proof which can be found in [3]. We call the collection of algebras satisfying these identities, the *binary hypercommutative (BH) variety*, and a member of this variety a *binary hypercommutative algebra (BHA).* We refer to identity (iii) as *binary hypercommutativity.* A second variety involves a ternary operation δ. This operation satisfies the identities:

(iv) $\delta(p,q,r)$ is invariant under any permutation of its three variables.
(v) $\delta(p,q, \delta(p,q,r)) = r$
(vi) $\delta(\delta(p,q,r), \delta(s,t,u), \delta(v,w,y))$ is invariant under any permutation of its nine variables.

We call the collection of algebras satisfying these identities, the *ternary hypercommutative (TH) variety,* and a member of this variety a *ternary hypercommutative algebra (THA).* We refer to identity (vi) as *ternary hypercommutativity.* In both varieties there is cancellation; for the BH variety if $a*b = a*c$ then $b = a*(a*b) = a*(a*c) = c$, and for the TH variety if $\delta(a,b,c) = \delta(a,b,d)$, then $d = \delta(a,b, \delta(a,b,d)) = \delta(a,b, \delta(a,b,c)) = c$. The significance of the TH variety can be seen in the following Theorem whose proof is given in [7].

Theorem 16 *Let γ be an irreducible cubic curve which inverts onto itself via a circle w with center X and define a ternary operation δ on the nonsingular points of γ by setting $\delta(a,b,c)$ equal to the the fourth point besides a,b,c which lies on circle (a,b,c) and on γ. Then (γ,δ) is a THA and $\delta(a,b,c) = X*\{(X*a)*(b*c)\}$.* □

The equation $\delta(a,b,c) = X*\{(X*a)*(b*c)\}$ is illustrated in Figure 13. In this Figure the cubic curve inverts onto itself through the orthogonal circles u, v with centers X, Y respectively. By Theorem 16 it must also be the case that $\delta(a,b,c) = Y*\{(Y*a)*(b*c)\}$ as illustrated. In case γ is a noncircular conic and we define a ternary operation δ on γ in the same way, we also obtain a THA. This fact is proved in [4]. In Chapter 8 we discuss in full circle systems with a noncircular conic envelope. The expression $\delta(a,a,c)$ represents the point besides a, c which lies on γ and on the circle through c and tangent to γ at a, and the expression δ(a,a,a) represents the point besides a which lies on γ and on the circle which intersects γ with multiplicity 3 at a. It is also possible that δ(a,a,a) = a, in which case there is a circle which meets γ with multiplicity 4 at point a. In this case we call the point a *δ-idempotent.* In case a (G,g) circle system Ω has a self-inversive cubic envelope λ, we say that Ω is *justified with respect to λ* in case the formula g - a - b - c = δ(a,b,c) = X*{(X*a)*(b*c)} holds for all points a,b,c of G, (even when two or more of the points a, b, c are identical). In this case Ω is a subalgebra of (λ,δ). For example, a $(Z_3,0)$ circle system contains only three points and thus contains no pure circles, so its points 0, 1, 2 can be put anywhere on λ, and it need not be justified.

Figure 13: Equation for δ(a,b,c)

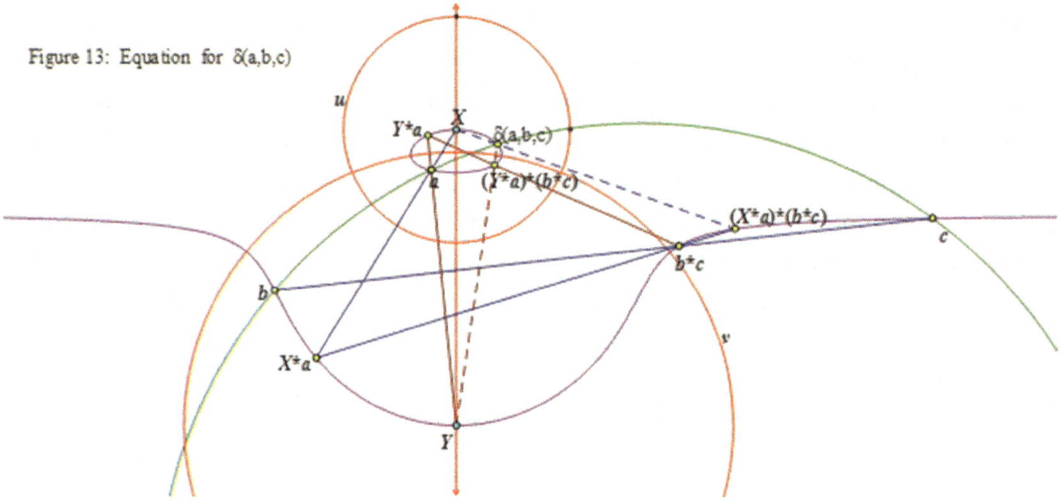

But if we require the $(Z_3,0)$ circle system to be justified on λ, then we must also satisfy δ(0,0,0) = δ(1,1,1) = δ(2,2,2) = 0, and δ(0,0,1) = δ(1,1,2) = 2. Since δ(0,0,0) = 0, the point 0 is δ-idempotent, and if we put 0 on the oval of a Type IV cubic, then there are four choices for 0 since there are exactly four δ-idempotent points on the oval which occur at intersections of inverters with the oval. We shall see in Section 7.1 that the equation δ(x,x,x) = p has exactly three solutions for any choice of p on λ, and thus once the position on the oval is chosen for vertex 0, the positions for vertices 1, 2, (on the oval), are forced. In Figure 14 we illustrate one of the four distinct justified $(Z_3,0)$ circle systems on the oval of a Type IV cubic.

Figure 14: justified $(Z_3,0)$ circle system on oval

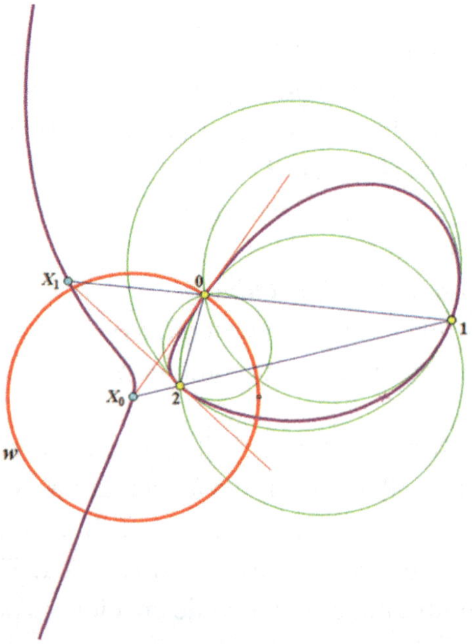

From this point on, whenever a circle system is considered along with its self-inversive cubic envelope, we shall consider that it is justified in relation to its envelope. It should be expected that if a circle system with a self-inversive cubic envelope has sufficiently many points, then it must be justified. This question will be adressed for circle systems with base Z_n in Section 4.2. We can make immediate use of the TH axioms to construct a (G,0) circle system on a cubic curve.

Theorem 17 *Let γ be an irreducible cubic curve which inverts onto itself via a circle w and has a flex point p at infinity. If Γ is a (G,0) triple system on γ with vertex 0 at point p, then Γ is also a (G,0) circle system .*

Proof Suppose p, q, r, s are points of Γ and p+q+r+s = 0. We will show that δ(p,q,r) = s. In case s, t, u is a collinear triple on γ, we have s+t+u = 0 and δ(s,t,u) = 0, the point at infinity on γ. Thus

(4) $\delta(s,t, -s-t) = 0.$

From (4) we immediately obtain:

(5) $\delta(s,t, 0) = -s-t$
(6) $\delta(0, t, -s-t) = s$
(7) $\delta(0, t, -t) = 0$ *and thus* $\delta(0, 0, -t) = t.$

Consider: $\delta(p,q,r) = \delta(\delta(0, q, -p-q), \delta(0,0,-q), \delta(0,0,-r))$ using (6), (7). Now using the ternary hypercommutativity of the THA (γ,δ) and (7) we have:

(8) $\delta(p,q,r) = \delta(\delta(0, q, -q), \delta(0,0,0), \delta(0,-r, -p-q)) = \delta(0, 0, \delta(0, -r, r+s))$

Then using (6), (7), (8) we obtain $\delta(p,q,r) = \delta(0, 0, -s)) = s.$ □

In case the abelian group G of Theorem 17 is Z or Z_n a procedure for constructing a (G,0) triple system on an irreducible cubic curve can be found in [2]. By Theorem 17 we then obtain a (G,0) circle system Γ although many of the *circles* actually consist of three collinear points and the point at infinity. If we want a *pure* circle system, i.e., one in which every four element set {a,b,c,d} with a+b+c+d = 0 corresponds to an ordinary circle, we need only invert by a circle v whose center does not lie on any of the *circles* of Γ. This, in fact, is how we constructed the (Z_{12},0) circle systems of Figure 10. In [5] a cubic which inverts onto itself is called a *circular cubic.* Also in [5] it is mentioned that the inverse of a circular cubic via a circle with center on the cubic is again a circular cubic, and the inverse by a circle with center off the cubic is a *bicircular quartic.* Thus in Figure 10, $\phi_v(\gamma)$ is a bicircular quartic and γ is a circular cubic. If γ is inverted by a circle u with center on γ but not on any of the points of the (Z_{12},0) triple system, then we obtain a pure (Z_{12},0) circle system on a circular cubic.

Chapter 3

Circle Systems with base Z

3.1 Equivalence classes and subsystems

In the previous Chapter we have seen the usefulness of the (Z,0) circle system for the construction of circle systems with base Z_n. This Chapter is devoted to a general consideration of circle systems with base Z. First we determine the equivalence classes of such systems. Let Ω be a (Z,0) circle system which we constructed in Theorem 8 and let $k \in Z$. By adding k to each vertex of Ω we obtain a (Z,4k) circle system. If Γ is a (Z,1) circle system and we add k to each vertex we obtain a (Z, 4k+1) circle system. Similarly, if we add k to a (Z,2) or (Z,3) circle system we obtain a (Z, 4k+2) or a (Z, 4k+3) circle system respectively. Thus every circle system with base Z is equivalent to (Z,0), (Z,1), (Z,2) or (Z,3) circle system. If we replace each point of a (Z,3) system with its inverse we obtain a (Z,-3) system, and if we further add one to each point we obtain a (Z,1) circle system. Thus every circle system with base Z is equivalent to a (Z,0), (Z,1) or (Z,2) system.

By Theorem 2, for a (Z,0) circle system there exists a circle w_0 which inverts every pair of points of the form {q,-q}, and by Theorem 4, vertex 0 must lie on w_0. There is no integer s satisfying s+s = 1 so a (Z,1) circle system has no circle complexes and so cannot be equivalent to a (Z,0) or a (Z,2) circle system. Recall that Ω_s denotes the collection of circle sets in a (Z_n, g) circle system Ω of the form $\{p_1, q_1, p_2, q_2\}$ where $p_1 + q_1 = p_2 + q_2 = s$, and s+s = g. A (Z,2) circle system contains the circle complex Ω_1 so there exists a circle w_1 which either inverts or antiinverts every pair of points {p,q} with p+q = 1, however w_1 contains no point of the circle system. We can thus conclude that the (Z,0), (Z,1) and (Z,2) circle systems are mutually nonequivalent. We have proved :

Theorem 18 *Every circle system with base Z is equivalent to either a (Z,0), (Z,1) or (Z,2) system, and these circle systems are mutually nonequivalent.* □

The (Z,0) circle system constructed in Theorem 8 actually contains within it both a (Z,1) and a (Z,2) circle system. Let Ω be a (G,g) circle system and if a, b, c $\in$ G let $\delta(a,b,c) = g - a - b - c$. It is easy to check that (Ω,δ) is a THA, (defined in Section 2.2.) Thus if a,b,c, δ(a,b,c) are distinct then these four points belong to a circle of Ω. If S is a subset of G which is closed under the ternary operation δ, then we call (S,δ) a *subalgebra* of (Ω,δ). Now let Ω be a (Z,0) circle system with $\delta(a,b,c) = -a-b-c$, and let $S = \{4k-1 : k \in Z\}$ and let $U = \{4k-2 : k \in Z\}$. It is then easy to see that (S,δ) and (U,δ) are subalgebras of (Ω,δ). Now relabel the points of S using the rule $4p-1 \rightarrow p$. Thus if a = 4p-1, b = 4q-1 and c = 4s-1 are points of S, then

$$\delta(a,b,c) = -(4p-1) - (4q-1) - (4s-1) = 4(-p-q-s+1) - 1$$

and our relabeling produces the four cocyclic points p, q, s, -p-q-s+1 which sum to 1. The result is that the points of S form a (Z,1) circle system. If we similarly relabel the points of U by the rule 4p-2 → p, we obtain a (Z,2) circle system. We say that the (Z,1) and (Z,2) circle systems occur as *subsystems* of the (Z,0) circle system. We have proved :

Theorem 19 *(Z,1) and (Z,2) circle systems exist and occur as subsystems of a (Z,0) circle system.* □

In Theorems 11, 13 we have shown that the vertices of a (Z,0) or (Z_n,0) circle system lie on an algebraic curve ξ which is the inverse of a self-inversive cubic. It follows from Theorem 16 and the fact that inversion maps *circles* to *circles*, that (ξ,δ) is a THA. Thus if Ω is a (Z,0) or (Z_n,0) circle system with envelope ξ, then (Ω,δ) is a subalgebra of (ξ, δ). The (Z,1) and (Z,2) circle systems of Theorem 19 thus occur on the self-inversive envelope of a (Z,0) system, but we have not as yet shown that these two circle systems *must* have such an envelope. Neither have we shown that these circle systems can only occur as subsystems of a (Z,0) circle system.

3.2 Construction of the (Z,2) Circle System

We will turn now to a direct construction of a (Z,2) circle system. To accomplish this we require the following three Lemmas whose proof can be found in [6]. In These Lemmas It must be allowed that the four points on a *circle* may, in some cases be collinear.

Lemma 20 *Any five of the circles (A,B,C,D), (E,F,G,H), (A,B,F,E), (B,C,G,F), (C,D,H,G), and (D,A,E,H) imply the sixth. See Figure 15.* □

Lemma 21 *If (A,B,D,F), (B,C,F,G), (C,A,E,G), and (E,F,D,G) are circles, then circles (A,B,C), (A,D,E) are tangent at A. Conversely, If circles (A,B,C), (A,D,E) are tangent at A, then any three of the circles (A,B,D,F), (B,C,F,G), (C,A,E,G), and (E,F,D,G) imply the fourth, See Figure 16.* □

Lemma 22 *Circles (C,D,E,F), (C,A,B,F), (D,A,B,E), (C,A,D), and (F,E,B) imply the existence of a circle we denote by (A,A,B,B) which is tangent to (C,A,D) at A and tangent to (F,E,B) at B. Conversely, If circle (A,A,B,B) is tangent to circle (C,A,D) at A and tangent to circle (F,E,B) at B, then any two of the circles (C,D,E,F), (C,A,B,F), (D,A,B,E) imply the third. See Figure 17.* □

Lemma 23 *If (A,B,E,F) and (C,D,E,F) are circles, and circles (A,B,C), (A,D,E) are tangent at A, then circles (A,D,C) and (B,F,C) are tangent at C. Conversely, if circles ((A,B,C), (A,D,E) are tangent at A, and circles (A,D,C) and (B,F,C) are tangent at C, then either one of the circles (A,B,E,F), (C,D,E,F) implies the other. See Figure 18.* □

For ease of notation we represent the circle configurations of Figures 15, 16, 17, 18 respectively as follows.

E			F	B			C	C			D	C			B
	A	B			A	A			A	A			A	A	
	D	C			D	E			B	B			D	E	
H			G	F			G	F			E	C			F

In this notation the four points at the outer corners as well as the four central points are cocyclic. The four points in the upper half are cocyclic as are the four points in the lower half of each diagram. The four points in the left half of each diagram represent four cocyclic points as do the four points in the right half. Thus each diagram represents a configuration consisting of six sets of four cocyclic points. In case one of these sets contain a repeated point then this repeated point belongs to two circles and represents a point of tangency between the two. Thus, for example, the repeated point A in the diagram for Figure 16 indicates that circles ABC and ADE are tangent at A.

Figure 15: Circle chain

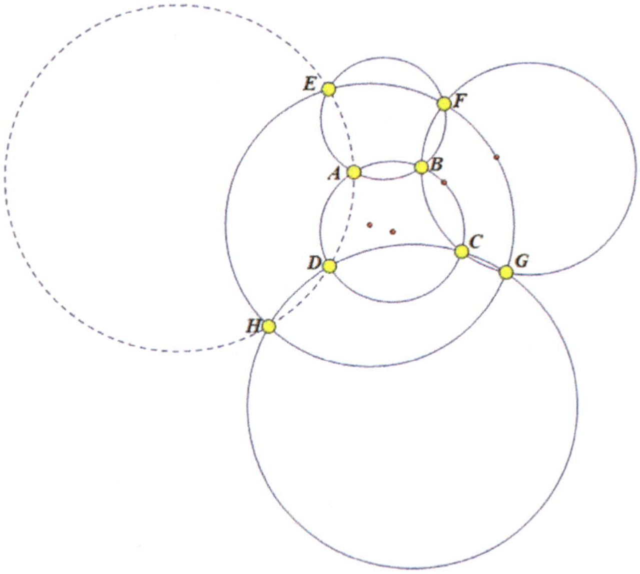

Figure 16: Circle chain variant

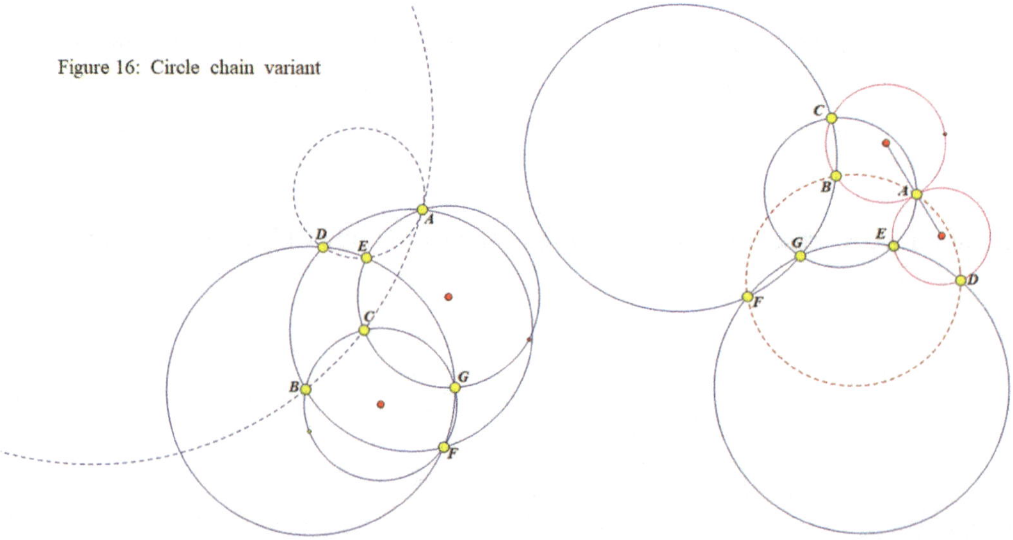

We will now present our construction of a (Z,2) circle system. If Ω is a (Z,2) circle system, then by Theorem 2 there is a circle w_1 which either inverts or antiinverts every pair of points {p,q} with p+q = 1. Both possibilities can occur. In our next Theorem we suppose w_1 is an inverter.

Theorem 24 *The following construction yields a (Z,2) circle system. Let* $w = w_1$ *be the circle which will invert all pairs {p,q} with p+q = 1, and let {0,1}, {-1,2}, {-2,3} be three such pairs as in Figure 19. Put vertex 4 on circle (1,-1,-2) and let* $-3 = \phi_w(4)$*, the inverse of point 4 with respect to w. Let vertex 5 be the contact point besides -3 of the circles (-3,1,-1), (-3,2,-2). Shift the generating vertices 1,2,3,4 until circle (0,-1,-2,5) is obtained. [This is not difficult to do if vertices 1,2,3,4 are chosen in the interior of circle w as in Figure 19.] Let* $-4 = \phi_w(5)$*, and let -5 be the contact point of circles (3,4,0),(5,2,0) different from 0. This is our starting configuration indicated by the bold circles in Figure 19. To complete the construction once vertices {0, ± 1, ..., ±(k − 1)} have been determined, let vertex* $k = \phi_w(-(k-1))$ *and vertex* $-k = (k,0,2)\cap(k,3,-1)$*.*

Figure 17: Circle chain variant

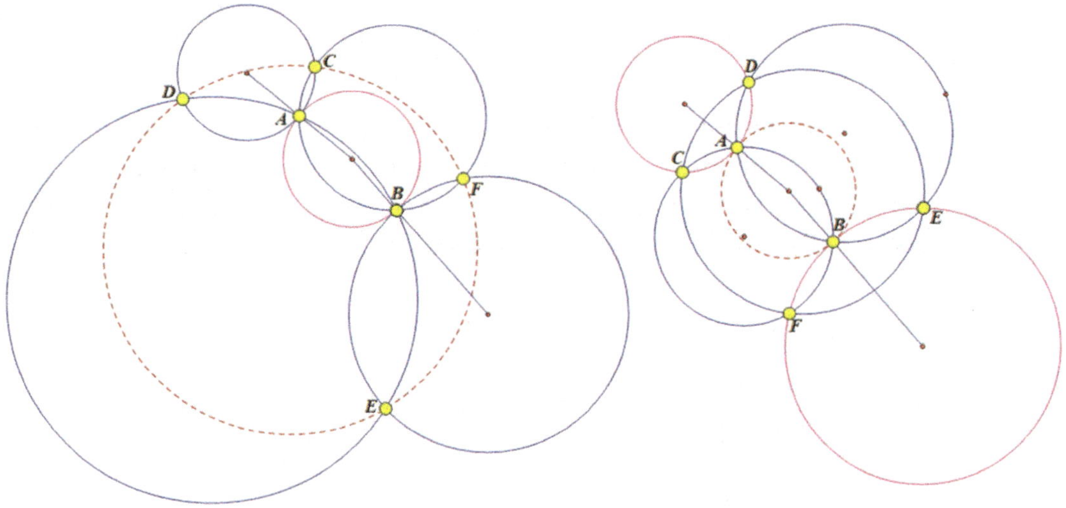

Figure 18: Circle chain variant

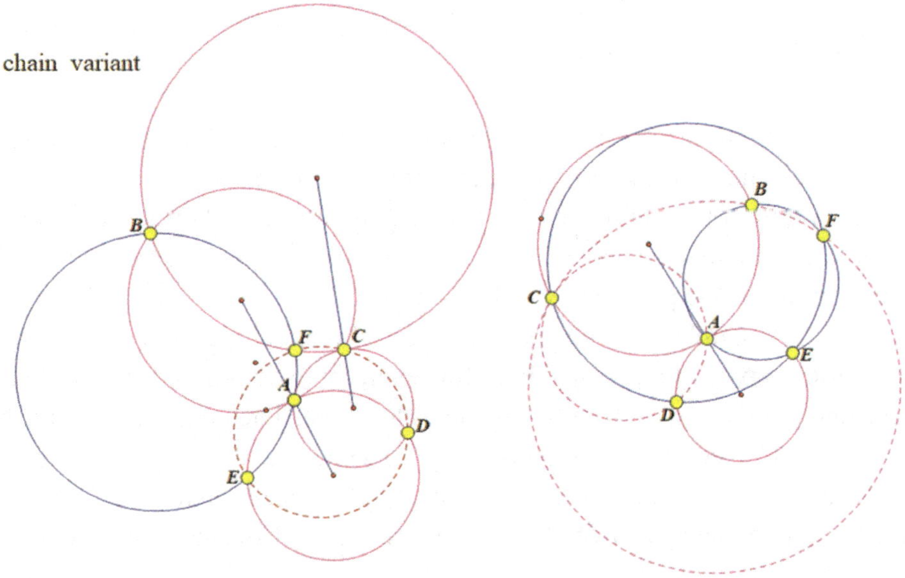

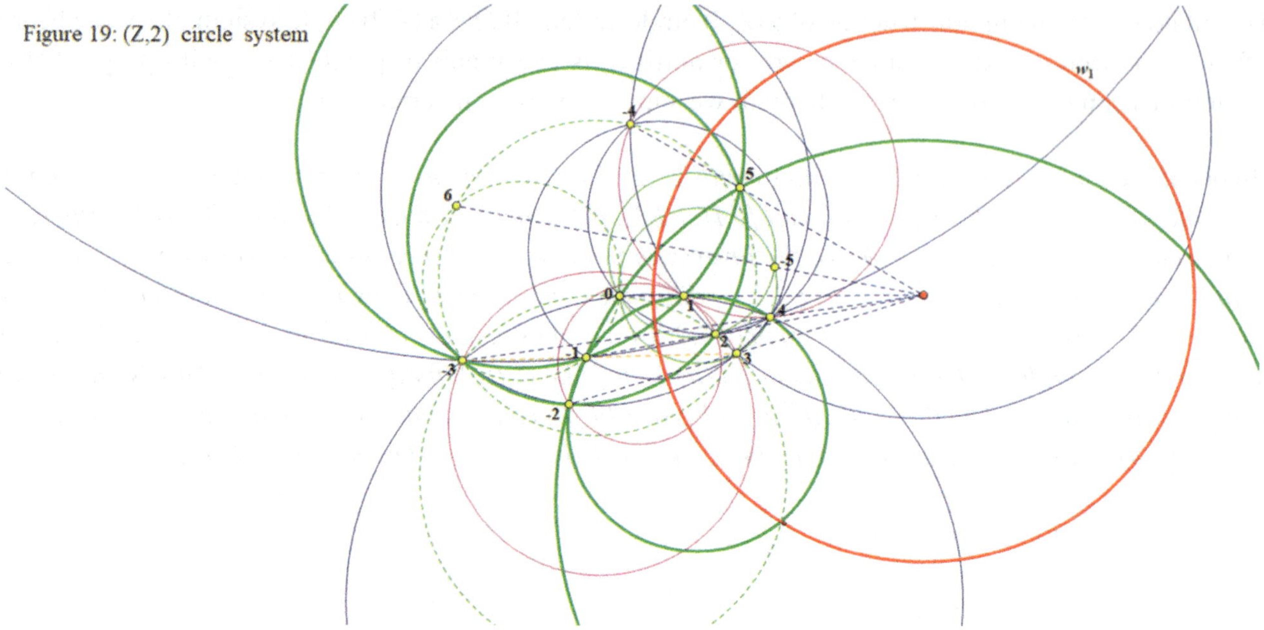
Figure 19: (Z,2) circle system

Proof Let Ω denote the set of integer vertices determined by our construction as well as all the circles with sum 2 so far obtained. One useful observation is that the inverse via w of a circle of Ω is again a circle of Ω, since if a+b+c+d = 2 then (1-a)+(1-b) + (1-c) + (1-d) = 2. Including the circles of our start configuration as well as their inverses we have the following imbalanced circles.

(4,-2,1,-1)	(-3,3,0,2)
(5,0,-1,-2)	(-4,1,2,3)
(5,2,-2,-3)	(-4,-1,3,4)
(5,1,-1,-3)	(-4,0,2,4)
(-5,4,3,0)	(6,-3,-2,1)
(-5,5,0,2)	(6,-4,1,-1)

Of course along with these we have all the balanced circles, i.e., circles of the form (p,q,r,s) with p+q = r+s = 1. If k is a positive integer we let S_k = {(q,a,b,c) with a+b+c+q = 2 and |a|, |b|. |c| all ≤ k and |q| = k}. We will show that Ω contains all circles over Z with sum 2 by proving all circles in S_k *as well as all tangencies* for each integer k. By *tangencies* we mean to show that if a+a+c +d = a+a+e+f = 2 then circles (a,c,d), (a,e,f) are tangent at point a. We proceed by induction on k. Note that S_1 contains no circles and S_2 = {(2,0,1,-1), (2,-2,1,1), (2,-2,2,0)}. Circles with one pair of repeated vertices we shall call *auxiliary*. In general, they must be listed in order to prove tangencies. In S_2 however, there do not exist two auxiliary circles with the same pair of repeated points so we have no tangencies to prove. The circle (2,0,1,-1) is balanced so we have it by our construction. Consider: S_3 = {(3,-3,0,2), (3,-3,1,1), (3,-2,1,0),(3,-1,0,0), (3,-1,2,-2), (3,3,-3,-1), (3,1,-1,-1)}. We have (3,-3,0,2) by construction, and we have the balanced circle (3,-2,1,0). To complete the proof of S_3 we must prove the tangency at 1 of (3,-3,1,1) and (2,-2,1,1). Consider the sequence of circle configurations:

1			-1	1			-1	-2			2
	6	4			6	4			1	1	
	-3	3			-2	2			-3	3	
-2			4	-3			5	6			4

The first configuration uses the balanced circle (3,-3,4,-2) and the start configuration circles (4,-2,1,-1), (6,-3,-2,1), (4,-4,3,-1), (6,-4,1,-1) to obtain circle (6,-4,3,-3) by Lemma 21. The second configuration uses the start configuration circles (5,-3,1,-1), (6,-3, -2,1), (6, -4 1,-1), (5,-3,2,-2) and the balanced circle (5,-4,2,-1) to obtain circle (6,-4,2,-2). The rightmost configuration uses the just proved circles (6,-4,3,-3) and (6,-4,2,-2) and the start configuration circles (6,-3,-2,1) and (-4,1,2,3) to obtain by Lemma 22 the desired tangency at point 1 of circles (3,-3,1,1), (2,-2,1,1). The circles in S_4 are:

c_2 = (-4,4,2,0)	(4,-3,1,0)	(4,-2,1,-1)
c_3 = (-4,4,3,-1)	(4,-3,2,-1)	
b_2 = (-4,3,2,1)	(4,-3,3,-2)	

The circles in the first column are all had by construction. The circles in the second column are balanced, so we also have these by construction. The remaining circle (4, -2,1,-1) has inverse (-3,3,0,2). Since we have this last circle by construction, we also have (4,-2,1,-1). Thus we have all circles in S_4. Quadruples in S_4 with repeated elements are:

(-4,4,1,1)	(4,0,0,-2)
(-4,-2,4,4)	(4,-1,-1,0)
(-4,3,3,0)	(4,-2,-2,2)
(4,-3,-3,4)	

For our argument we need not be concerned with quadruples with two pairs of repeated elements or which have an element repeated three times. Also we need not be concerned with the auxiliary circle (-4,-2,4,4) since no auxilliary circle in S_2 or S_3 can be tangent to it. Tangencies involving auxiliary circles in S_4 which do not contain -4 are all obtained by taking inverses since the inverses will form a tangency in S_3. For example suppose we want to prove that (4,-1,-1,0) is tangent to (3,-1,-1,1) at -1. A circle configuration to prove this would be given by

4			0	-3			1
	-1	1			2	2	
	1	3			0	2	
-2			0	3			1

The configuration on the right is the inverse of the configuration on the left. The circles and tangencies in the right configuration have been proved since no entry has absolute value greater than 3. Since inversion preserves tangencies we can conclude that our original tangency is proved. We must show that (-4,4,1,1) is tangent at vertex 1 to both (2,-2,1,1) and (3,-3,1,1). The relevant circle configurations are:

-4			4	-4			4
	1	1			1	1	
	2	-2			3	3	
3			-1	2			0

In both configurations all the required component circles have been proved, namely (-4,1,2,3), (3,2,-2,-1), (4,-2,1,-1), (4,-4,3,-1), (3,-3,2,0), (4,-4,2,0), (4,-3,1,0) and thus these tangencies are obtained by Lemma 21. Actually it is only necessary to prove one tangency involving (-4,4,1,1), (or any particular auxiliary circle in S_k since) since the remaining tangencies will follow by transitivity. For example, in S_3 we know that auxiliary circles (3,-3,1,1), (2,-2,1,1) are tangent at vertex 1. So if (4,-4, 1,1) is shown to be tangent at vertex 1 to (2,-2,1,1), then it follows by transitivity that (4,-4,1,1) is tangent to (3,-3,1,1) at vertex 1. The only remaining tangency is proved using the circle configuration:

-4			0
	3	3	
	-1	-3	
4			2

And thus all circles and all tangencies for S_4 are proved. Now assume inductively that all circles in S_t with $4 \leq t < k$ have been proved along with all tangencies involving such circles. We will prove all circles in S_k including all tangencies which involve a circle in S_k and a circle in S_t, $t < k$. Let $U_k = \{c_p = (-k,k,p,2-p) : 0<p<k\}$; let $V_k = \{b_p = (-k, k-1,p, 3-p): 2\leq p\leq k-2\}$ and let $W_k = \{(-k, a,b,c): |a|,|b|,|c|\leq k-2\}$. These three sets constitute all the circles in S_k which contain -k. Circles in S_k which do not contain -k are either balanced or inverses of circles in S_{k-1} which we have by the inductive hypothesis. We first show that every circle in U_k is obtained. Note that, by virtue of our construction, we have circles $c_2 = (-k,k,2,0)$ and $c_3 = (-k,k,3,-1)$. The circle configuration:

k			2
	-k	0	
	3	k-1	
-1			1-k

Uses circles c_2, c_3 along with the balanced circle (2,-1,k,1-k) and the two circles (k-1, 1-k,0,2), (k-1,1-k,3,-1) which we have by the inductive hypothesis, to obtain circle b_3 = (-k,k-1,3,0). Next consider the circle configuration:

0			2
	-k	k	
	k-1	3-k	
3			-3

Which uses c_2, b_3; the start configuration circle (3,-3,0,2); the circle (k-1,3-k,2,-3) obtained by the inductive hypothesis, and the circle (k, 3-k,-3,2) which is obtained as the inverse of the circle (-k+1,k-2,4,-1) which is covered by our inductive hypothesis, to obtain circle c_{k-1} = (-k,k,k-1,3-k). Now assuming that circle c_p has been proved for $3 \le p \le k-2$, consider the circle configuration:

k			3-k
	-k	k-1	
	-p+2	p+1	
p			-p-1

which uses circles c_p, c_{k-1}. If $p<k-2$ then (p+1, -p-1, p,-p+2), (k-1,3-k,p+1,-p-1) are circles which we have by the inductive hypothesis. Circle (k, 3-k,p,-p-1) is the inverse of β = (-k+1, k-2, -p+1, p+2) which we have by the inductive hypothesis. If p = k-2 our configuration has the form:

k			3-k
	-k	k-1	
	4-k	k-1	
k-2			1-k

The outer circle (β) is balanced and the upper circle is c_{k-1} so we have these. The bottom circle (k-1, 1-k,4-k,k-2) we have by the inductive hypothesis. The tangency (3-k,1-k,k-1,k-1), (-k,4-k,k-1,k-1) at k-1 can be obtained by considering the circle configuration:

-k			4-k
	k-1	k-1	
	3-k	1-k	
k			k-2

Here the left circle is c_{k-1}, the bottom circle is balanced and the right circle we have by the inductive hypothesis. So the above configuration proves the circle b_{p+1} = (-k, k-1, p+1, -p+2). We have shown that circles c_p, c_{k-1} imply circle b_{p+1}. Finally consider the circle configuration:

k-1			3-k
	-k	k	
	p+1	-p+1	
-p+2			p-2

Which uses circles c_{k-1}, b_{p+1} ; circle (k,3-k,-p+1,p-2) which is the inverse of (-k+1,k-2,p, -p+3) which we have by the inductive hypothesis; circle (p+1, -p+1,-p+2,p-2) which we have by the inductive hypothesis, and circle (k, 3-k,-p+1,p-2) which is the inverse of circle (-k+1,k-2,p,-p+3) which we also have by the inductive hypothesis. So this configuration proves, by Lemma 20, the circle c_{p+1} = (-k,k,p+1,-p+1). At this point every circle in U_k has been proved and along the way every circle in V_k except for b_2 = (-k,k-1,2,1) has been proved. This last circle is proved using the circle configuration:

k			0
	-k	2	
	k-1	1	
3-k			-1

Which uses c_2, c_{k-1}, (0,2,1,-1), (3-k,k-1,1,-1) which is obtained by the inductive hypothesis and (k,3-k,0,-1) which is the inverse of (-k+1,k-2,1,2) which we also have by the inductive hypothesis, to obtain b_2. Next we prove the circles in W_k. We assume inductively that all circles (a,b,c,d) with a+b+c=d = 2 and |a|, |b|, |c|, |d| < k, as well as all tangencies involving such circles have been proved. Let (-k,a,b,c) be a typical quadruple in W_k with $k-2 \geq a > b > c$ and b>0, and consider the circle configuration :

k			2-a
	-k	a	
	b	c	
2-b			-c

In this configuration c_a = (-k,k,a,2-a) and c_b = (-k,k,b,2-b) lie in U_k and so have been proved. The quadruples (b,2-b,c,-c), (a,2-a,c,-c) have all entries with absolute value less than k so they are either circles or involve tangencies which we have by the inductive hypothesis. The inverse of the outer circle, (in case it is a circle), is (-k+1, a-1, b-1, c+1) which we have by the inductive hypothesis. In case it is an auxiliary circle, for example if -c = 2-a, then the tangency at -c of (a,c,-c,-c) and (k,2-b,-c,-c) is proved by the tangency of the inverses. All the circles in W_k are thus proved, and with these we have proved all circles in S_k. It remains to prove tangencies which involve circles in S_k and circles in S_t for t<k. We need only consider such tangencies which involve auxiliary circles of the form (-k, t, s,s) with $t \neq s$, |t|<k-1 and $s \in \{1,2, \ldots, k-1\}$ since tangencies involving auxilliary circles of this form without -k can be obtained by taking inverses and applying the inductive hypothesis. Consider the circle configuration:

k			2-s
	-k	s	
	t	s	
2-t			-s

which employs proven circles c_s = (-k,k,s,2-s); c_t = (-k,k,t,2-t); the circle (t,2-t,s,-s) which we have by the inductive hypothesis, and the circle (k, 2-s,2-t,-s). If s = k-1 then this last circle is balanced, and if s<k-1 then we have the inverse (-k+1, s-1,t-1,s+1) by the inductive hypothesis. Thus the above configuration proves the tangency at s of the auxiliary circles (-k,t,s,s) and (2-s,-s,s,s). Now if (a,b,s,s) is a quadruple with a+b+2s = 2; b≠-s and |a|, |b| < k, then we have the tangency at s of the auxiliary circles (a,b,s,s) and (2-s,-s,s,s) by the inductive hypothesis. We can conclude by transitivity that (-k,t,s,s) is also tangent to (a,b,s,s) at s. With all circle and all tangencies of S_k proved, we can conclude by induction that Ω is a (Z,2) circle system. □

There also exists a (Z,2) circle system with a circle w_1 which antiinverts every pair {p,q} with p+q = 1. The construction and proof is the same as in Theorem 24 except that antiinversion is used instead of inversion to define points. The start configuration is indicated by the bold circles in Figure 20. If our inversive (Z,2) circle system Ω is inverted by a circle v with its center on w_1, then w_1 inverts to a line L and each pair {p,q} of points with p+q = 1 are reflections of each other across this line. If we invert Ω by a circle with its center at vertex 0, then we obtain a (Z,2) geometric triple system which has an irreducible cubic envelope γ by Theorem 10. As a consequence Ω has an envelope which is an algebraic curve ξ which is the inverse of γ. Thus after inversion via v, we obtain an algebraic curve $\phi_v(\xi)$ which contains infinitely many pairs of points which are reflections of each other across L. We can then deduce by an argument similar to that used in Lemma 12, that $\phi_v(\xi)$ is itself symmetric with respect to L, and thus that ξ inverts onto itself via the circle $\phi_v(L) = w_1$. As a consequence of Theorem 14, there is a second circle, orthogonal to w_1 which also inverts ξ onto itself. In case Ω is a (Z,2) circle system which antiinverts onto itself via a circle w_1 the results developed in the next Section will be useful.

Figure 20: antiinversive (Z,2) circle system

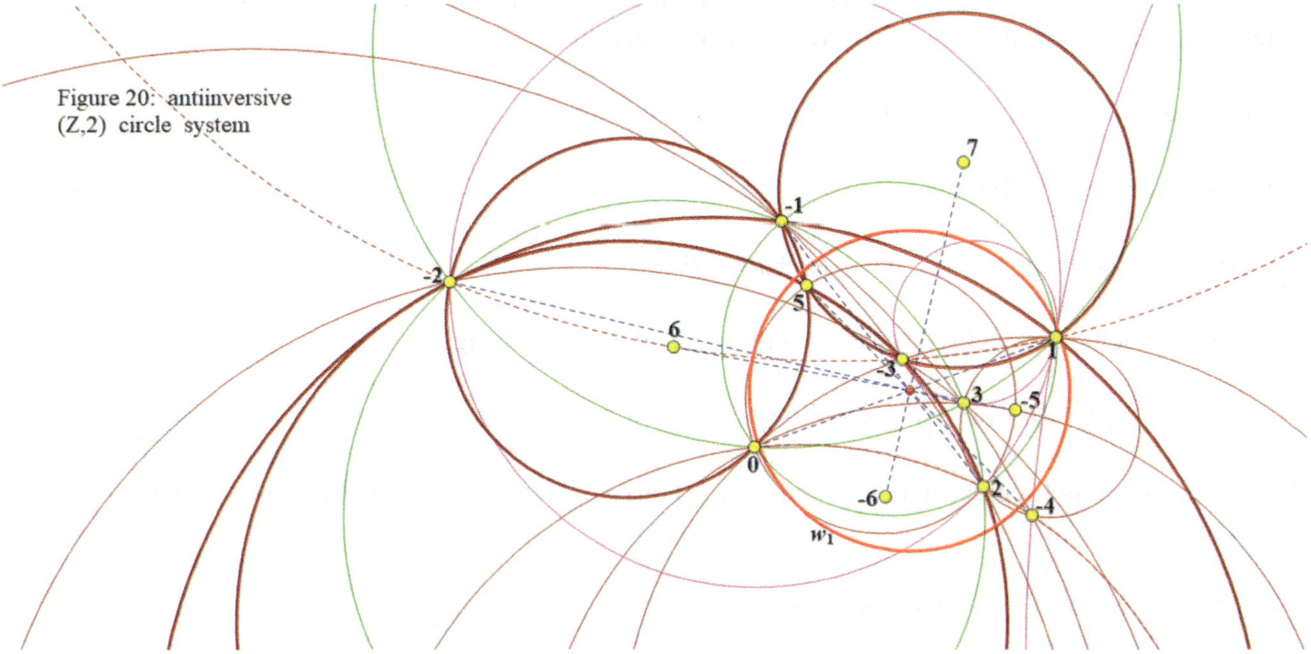

3.3 Properties of Cubic Curves which Antiinvert onto Themselves

A Type V cubic which inverts onto itself, inverts onto itself through two mutually orthogonal circles, and a Type IV cubic which inverts onto itself, inverts onto itself through three mutually orthogonal circles, and antiinverts onto itself through a fourth circle. In Figure 21 these four circles are depicted in relation to a Type IV cubic. These results can be found in [7] and [5]. In this Section we obtain similar results if we assume first that γ is an irreducible cubic which *antiinverts* onto itself. This is accomplished by considering the equation of γ.

Lemma 25 *The equation of an irreducible cubic curve which antiinverts onto itself can be put in the form:* $x(x^2 + y^2 - 1) + (ex^2 + fxy + gy^2) = 0$.

Proof Let γ be an irreducible cubic curve which antiinverts onto itself via a circle w. We can choose a coordinate system so that w is the unit circle and γ has a vertical asymptote. As a consequence the origin must lie on γ and the point (0,1,0) at infinity on vertical lines must lie on the homogenized equation of γ. Then the equation of γ has the form

(1) $ax^3 + bx^2y + cxy^2 + ex^2 + fxy + gy^2 + hx + ky = 0.$

Antiinversion of γ is accomplished by replacing x, y in (1) by -x/M and -y/M respectfully, where $M = x^2 + y^2$. When this is done and fractions are cleared we obtain

(2) $-ax^3 - bx^2y - cxy^2 + M(ex^2 + fxy + gy^2) - M^2(hx + ky) = 0.$

Since γ is to antiinvert onto itself, equations (1), (2) must represent the same curve. The terms in (2) with M as a factor have degree greater than 3, but if these are 0 we are left with $-ax^3 - bx^2y - cxy^2$ which is not irreducible. It must then be that M is a factor of $-ax^3 - bx^2y - cxy^2$ thus allowing us to cancel a factor of M from each term of (2) and obtain a cubic. Long dividing $-ax^3 - bx^2y - cxy^2$ by M we obtain a quotient of -ax and a remainder of $-bx^2y + (a-c)xy^2$. The remainder must be identically zero yielding b = 0 and c = a. Then after removing the common factor of M from (2) we obtain

(3) $-ax + (ex^2 + fxy + gy^2) - (x^2 + y^2)(hx+ky) = 0.$

Since b = 0 and c = a, (3) reduces to

(4) $ax^3 + axy^2 + ex^2 + fxy + gy^2 + hx + ky = 0.$

Comparing coefficients in (3), (4) we obtain h = -a and k = 0 yielding

(5) $ax(x^2 + y^2 - 1) + ex^2 + fxy + gy^2 = 0.$

Equation (5) will not represent a cubic unless a is nonzero, so we can divide by a and rename coefficients to obtain

(6) $x(x^2 + y^2 - 1) + ex^2 + fxy + gy^2 = 0$

as the final equation of γ. □

In [7] it is shown that the equation of an irreducible cubic curve ξ which inverts onto itself via the unit circle is given by

$$(7)\ (ax + by)(x^2 + y^2 + 1) + ex^2 + fxy + gy^2 = 0.$$

We can simplify this further by requiring that ξ have a vertical asymptote. Then the point (0,1,0) must lie on the homogenized form of (7) and (7) reduces to

$$(8)\ ax\,(x^2 + y^2 + 1) + ex^2 + fxy + gy^2 = 0.$$

Again, we cannot have a = 0 so we can divide by a and rename coefficients to obtain

$$(9)\ x\,(x^2 + y^2 + 1) + ex^2 + fxy + gy^2 = 0.$$

The close similarity of equations (6), (9) prompts us to attempt to transform one equation into the other by a translation and a dilation. This will allow us to prove

Theorem 26 *Let γ be an irreducible cubic curve which antiinverts onto itself via a circle w. Then there exist three mutually orthogonal circles each of which invert γ onto itself.*

Proof We can suppose that the equation of γ is given by (6). To obtain a translation of γ s units to the right and t units up we replace x, y in (6) by (x+s), (y+t) respectively to obtain

$$(10)\ (x+s)((x+s)^2+(y+t)^2 -1) + e(x+s)^2+f(x+s)(y+t)+g(y+t)^2 = 0.$$

We want this translated curve to contain the origin so the constant term in (10) must be zero, yielding

$$(11)\ s(s^2+t^2-1) + es^2 + fst + gt^2 = 0.$$

Note that this is the equation obtained by requiring the point (s,t) to lie on the equation (6) of γ. Then (10) reduces to

$$(12)\ x(x^2 + 2sx + y^2 + 2ty + s^2+t^2-1) + s(x^2 + 2sx + y^2 + 2ty)$$

$$+ex^2 + 2sex +fxy + ftx + fsy + gy^2 + 2tgy = 0.$$

Now to effect a dilation centered at the origin, we replace x, y in (12) by rx, ry respectfully. Then dividing by r^3 we obtain

$$(13)\ x^3 + xy^2 + x^2\left(\frac{3s+e}{r}\right) + xy\left(\frac{f+2t}{r}\right) + x\left(\frac{3s^2+t^2+2se+ft}{r^2}\right)$$

$$+ y\left(\frac{2st+fs+2gt}{r^2}\right) + y^2\left(\frac{g+s}{r}\right) = 0$$

In (13) we really don't care about the coefficients of x^2, y^2 or xy, but the coefficients of x, y must match the coefficients of x, y in (6) yielding

(14) $\left(\frac{3s^2+t^2-1+2se+ft}{r^2}\right) = 1$ *and* $\left(\frac{2s+fs+2gt}{r^2}\right) = 0$, or

(15) $3s^2+t^2-1+2se+ft=r^2$ = and

(16) $2^{st}+fs+2gt=0$

We note that (16) can be obtained by differentiating (6) with respect to y and setting $\frac{dx}{dy}=0$. This is so since we must translate to the origin a point of γ with vertical tangent. If we multiply (16) by t/2 and subtract the result from (11) we obtain

(17) $s^3 - s + es^2 + \frac{fst}{2} = 0.$

Factoring out s and multiplying by 2 we obtain

(18) $2s^2 - 2 + 2es + ft = 0.$

Now solving (18) for 2es + ft and substituting for this expression in (15) we obtain the nice result

(19) $r^2 = s^2 + t^2 + 1.$

This formula gives the radius r of a circle with center (s,t) which inverts γ onto itself. Solving (18) and (16) and eliminating t we obtain the cubic

(20) $4s^3 + (4e+4g)s^2 + (-4 + 4eg - f^2)s - 4g = 0$

Which either has one or three real roots for s. As a consequence γ inverts onto itself either through one circle or through three circles. However, by Theorem 14, if γ inverts onto itself via one circle w then it also inverts into itself via a second circle orthogonal to w. Thus (20) must have three real solutions and γ must invert onto itself through three circles. In [7], [5] it is shown that these three circles are mutually orthogonal. □

In Figure 21 we illustrate a cubic curve which antiinverts onto itself via a circle w, and inverts onto itself through three mutually orthogonal circles.

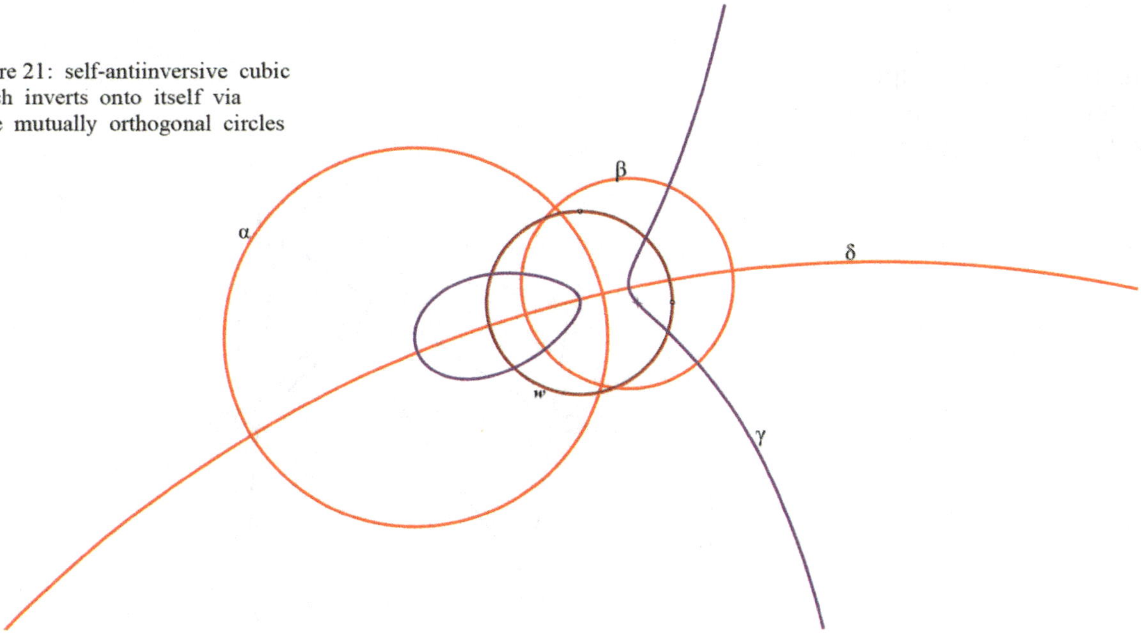

Figure 21: self-antiinversive cubic which inverts onto itself via three mutually orthogonal circles

As a consequence of Theorem 26, we can conclude, in particular, that a (Z,2) circle system whose envelope antiinverts onto itself also has the property that there exist three mutually orthogonal circles which invert the envelope onto itself. The construction of a (Z,1) circle system will be given in the next Section.

3.4 The (Z,1) circle system

The points of a (Z,1) circle system do not invert onto themselves through any circle, nevertheless we have the following perfect construction. We actually construct a (Z,1) triple system which is also a (Z,1) circle system, albeit with *circles* containing vertex 0 consisting of three collinear points with sum 1 and the point 0 at infinity. We can then obtain a pure (Z,1) circle system by inverting by a circle whose center is not a vertex.

Theorem 27 *The following constuction yields a (Z,1) geometric triple system which is also a (Z,1) circle system. (i)Put the pairs of points {2,-1} and {3,-2} on parallel lines so that vertex 0 will be at infinity. (ii) Let vertex 1 be the point of intersection, besides vertex 2, of line [2,-2] and circle (3,-1,-2). (iii) let vertex -3 be the point of intersection, besides vertex 3, of line [1,3] and circle (3,2,-1). (iv) let vertex 4 denote the intersection of lines [-1,-2], [-3,0] and let vertex -4 be the intersection of lines [1,4], [2,3]. (v) for k ≥5 define the points recursively, by setting vertex k = [-1, -k+2]∩[0, -k+1] and vertex - k = [1,k]∩[2, k-1].*

Proof The starting configuration for our construction is indicated by the bold lines and circles in Figure 22.

Figure 22: (Z,1) triple system which is also a (Z,1) circle system

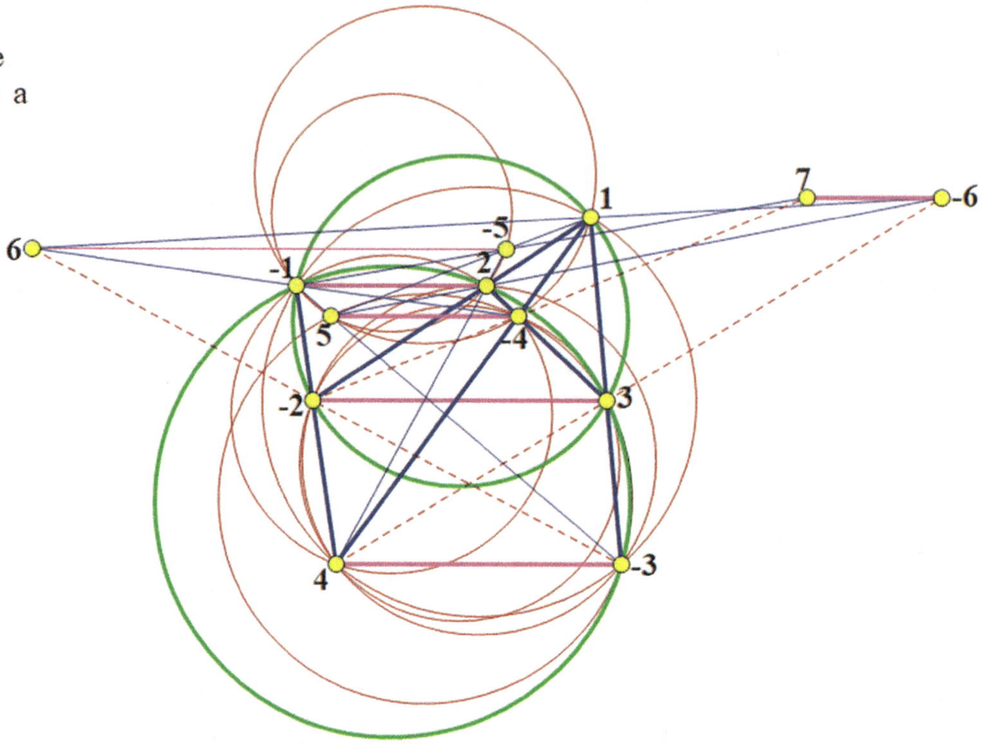

It is shown in [2] that this same starting configuration along with the definition of points given in the Theorem statement yields a (Z,1) triple system. Here we show by induction that Ω is also a (Z,1) circle system. Let S_k denote the set of all subsets {a,b,c,d} of Z with a+b+c+d = 1 and |a|,|b|,|c|,|d|≤k. For the basis we will show that all the four-tuples in S_4 are cocylic. Note that, since we have shown that Ω is a (Z,1) triple system, we have all *circles* (a,b,c,0) with a+b+c = 1; these consist of three collinear points and the point 0 at infinity. When, at the end, we invert our construction via a circle with a nonvertex center p, these *circles* will become ordinary circles with vertex 0 at p. Also, triples of the form {p,q,0} with p+q = 1 form mutually parallel lines. These parallel lines turn into circles mutually tangent at vertex 0 after our last inversion. The circles (1,-1,3, -2), (3,-3,2,-1) we have by construction, and circles (2, -2,1,0), (3,-3,1,0), (4,-4,1,0) we have since they contain vertex 0. The remaining circles in S_4 are:

(4,-4,2,-1)

(4,-4,3,-2)

(4,-3,1,-1)

(4,-3,2,-2)

(4,-1,-2,0)

(-4,3,2,0)

of which we need only prove the ones without vertex 0. The circle configuration

0			0
	4	-3	
	-1	1	
-2			3

proves circle (4,-3,1,-1). Note that we use the tangency of the auxiliary circles (4, -3,0,0), (3,-2,0,0) at vertex 0. Then the circle configuration

0			1
	-4	4	
	2	-1	
3			-3

Proves circle (4,-4,2,-1), and the circle configuration

2			-1
	-4	4	
	3	-2	
0			0

proves circle (-4,4,3,-2). Finally, the circle configurations

0			-1
	1	1	
	2	-3	
-2			4

-1			1
	4	-3	
	-2	2	
0			1

prove the remaining circle (4,-3,2,-2). The first configuration proves the tangency at vertex 1 of the auxiliary circles (0,-1,1,1), (2,-3,1,1), which is then used in the second configuration. Now suppose $k>5$ and that all circles in S_{k-1} have been proved. To complete the induction we prove the circles in S_k. These we partition into four classes: class A contains all circles of the form $\{A_t = (-k, k,-t, t+1), t = 1,2, \ldots, k-2\}$; class B contains all circles of the form $\{B_t =(k, -k+1, t, -t), t = 1,2, \ldots, k-2\}$; class C consists of circles of the form $\{(k, -s, -t, q), q<t<s\leq k-2, q,t,s \neq 0\}$; class D consists of circles of the form $\{(-k, s, t, q), k-2 \geq s, t > q, s,t,q \neq 0\}$, and class E consists of circles in S_k with a 0 component. As per our discussion above, we already have all class E circles. First suppose k is even with $k \neq 2t$ and consider the circle configuration

0			0
	k	-k+1	
	-t	t	
t-k+1			k-t

which proves circle B_t. Note that circle (-t, t, t-k+1, k-t) we have by the inductive hypothesis; the two side circles (0, k, -t, t-k+1), (0, -k+1, t, k-t) we have since they have a 0 component, and the auxiliary circles (k, -k+1, 0, 0), (t-k+1, k-t, 0, 0) we have by virtue of our construction. In case k = 2t our circle configuration contains a tangency which needs to be proved.

0			0
	k	-k+1	
	-t	t	
-t+1			t.

The circle configuration:

-k+1			0
	t	t	
	1-t	-t	
k-1			1

proves the required tangency of the auxiliary circles (-k+1, 0, t, t), (1-t, t, 0, 0) at point t, and thus all class B circles are proved. Now to prove class C circles consider the circle configuration

-k+1			s
	k	-s	
	-t	q	
t			1-q

with $t \neq 1-q$. The left side and top circles are B_t, B_s respectively which we have just proved. The right, bottom and outer circles we have by the inductive hypothesis, so this configuration proves the class C circle (k, -s, -t, q). In case t = 1-q we can interchange the positions of q and -t to avoid the tangency

-k+1			s
	k	-s	
	q	-t	
-q			t+1

and thus obtain the remaining class C circle. Now for $k \neq 2t$ consider the circle configuration

0			1
	-k	k	
	t+1	-t	
k-t			t-k

which proves circle A_t. Note that the right circle (k, -t, t-k, 1) is class C, just proved. In case k = 2t this configuration contains a tangency

0			1
	-k	k	
	t+1	-t	
t			-t.

To prove the tangency we use the circle configuration:

t+1			t
	-t	-t	
	1	k	
-1			1-k.

This uses the proved circles B_1 = (k, 1-k, 1, -1) and B_t = (k, 1-k, t, -t) to prove the required tangency of the auxiliary circles (t+1, t, -t, -t), (k, 1, -t, -t) at -t. We have now proved all class A circles. Finally, to prove class D circles consider the circle configuration

k			1-s
	-k	s	
	t	q	
1-t			-q

With $t \neq q+1$. The top circle is A_{s-1}, and the left side circle is A_{t-1}, just proved. The outer circle is class C which we have also proved, and the right side and bottom circles we have by the inductive hypothesis. So this configuration proves the class D circle (-k, s, t, q). In case t = q+1, we have a tangency at 1-t which is proved by the same circle configuration. All class D circles are thus proved, and we can conclude by induction that Ω is a (Z,1) circle system. □

It is evidently true that the cubic envelope of a (Z,1) circle system is self-inversive but we have not been able to find a proof so we leave it as a Conjecture.

Conjecture 28 *The points of a (Z,1) circle system lie on an irreducible cubic curve which is self-inversive or on the inverse of such a curve.*

Chapter 4

Perfect Polygons and Circle Systems

4.1 (Z_n,1) and (Z_n,2) Circle Systems

Recall that a *subsystem* of a circle system Ω is a subset of the vertex set of Ω which, after relabeling, forms a circle system in its own right. This concept was inroduced in Theorm 18 of Chapter 2, where it is shown that a (Z,0) circle system contains both a (Z,1) and a (Z,2) subsystem. We begin this Chapter by arguing similarly that a (Z_{4t},0) circle system contains a (Z_t,1) and a (Z_t,2) subsystem.

If Ω is a (Z_n,0) circle system and n = 4t then we can find (Z_t,1) and (Z_t,2) circle systems as subsystems of Ω. To do this we partition Z_n into the four cosets: ⟨4⟩, -1+ ⟨4⟩, -2+ ⟨4⟩, -3+ ⟨4⟩. The relabelings 4k→k; 4k-1→k; 4k-2 →k and 4k-3 →k turn these cosets to (Z_t,0), (Z_t,1), (Z_t,2), and (Z_t,3) circle system respectively. The further relabeling k →-k+1 converts the (Z_t,3) circle system into a (Z_t,1) circle system. So by Theorem 6 we obtain every possible circle system with base Z_n as a subsystem of Ω. In Figure 23 we have constructed a (Z_{32},0) circle system Ω using the method described in Theorem 9. The generating vertices, in particular vertex 4, must be shifted until vertex 16 lies on the inverting circle w_0. It is a difficult construction since the location of vertex 16 changes radically with even a minute shift of vertex 4. In Figure 24 we illustrate the (Z_8,1) subsystem, and in Figure 25, the (Z_8,2) subsystem of Ω.

Figure 23: $(Z_{32},0)$ circle system

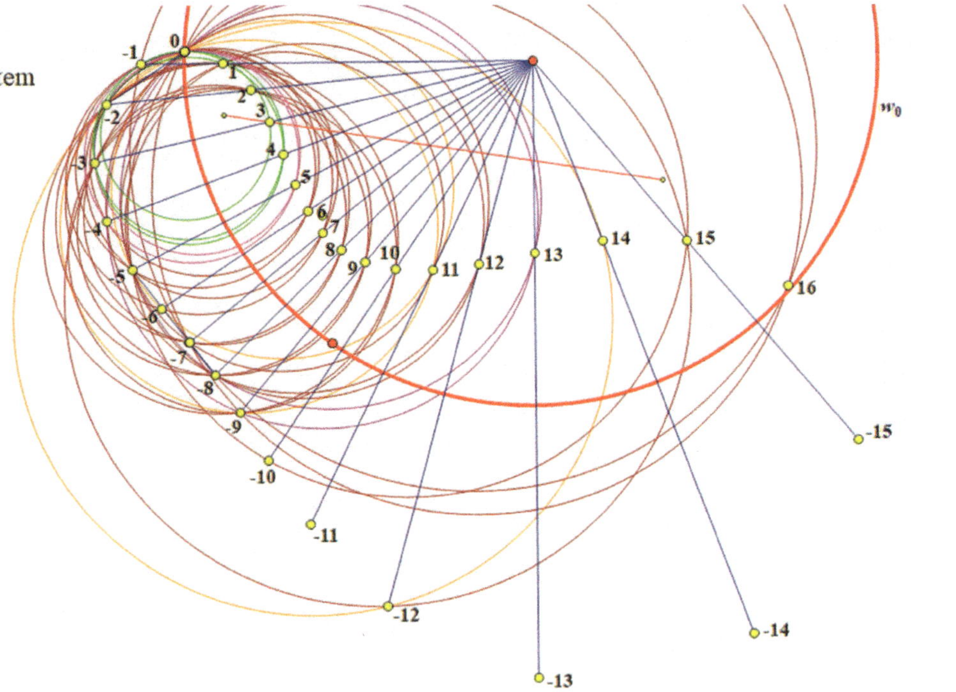

Figure 24: $(Z_8,1)$ circle system realized as subsystem of $(Z_{32},0)$

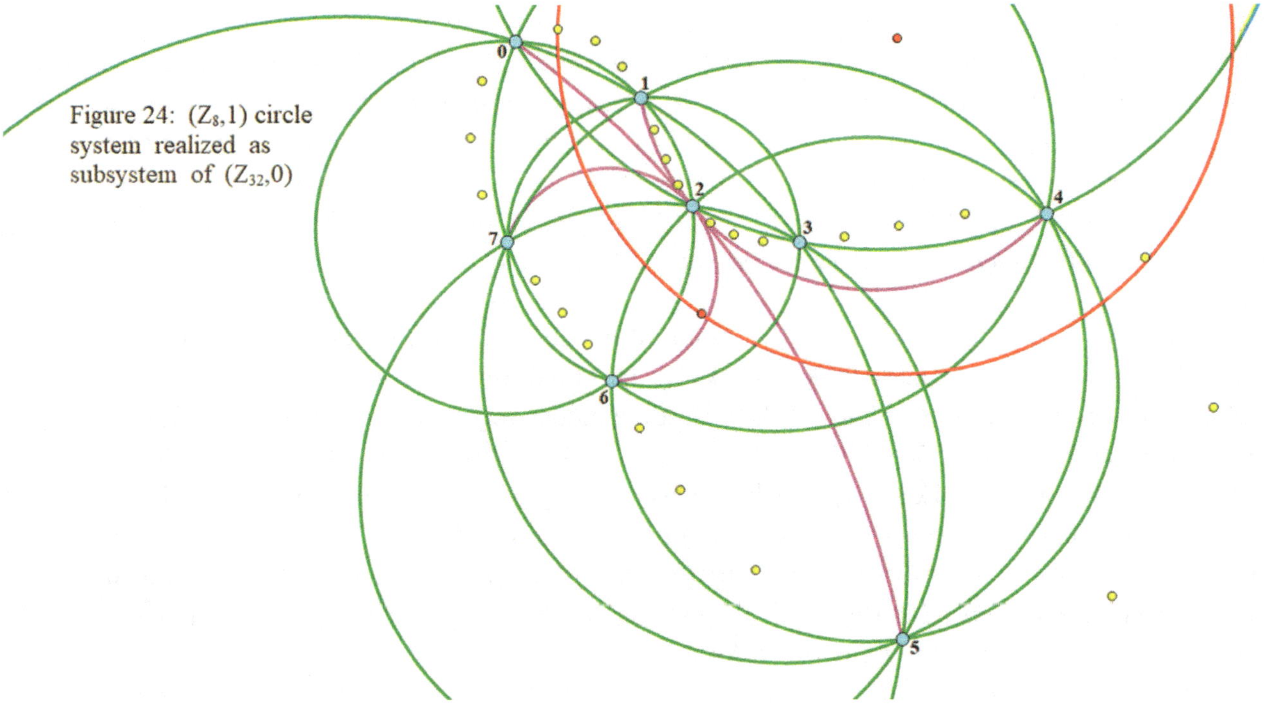

An easier construction is obtained if we start with a self-inversive cubic curve and construct a $(Z_{32}, 0)$ circle system on it by following the prescription given in our next Theorem.

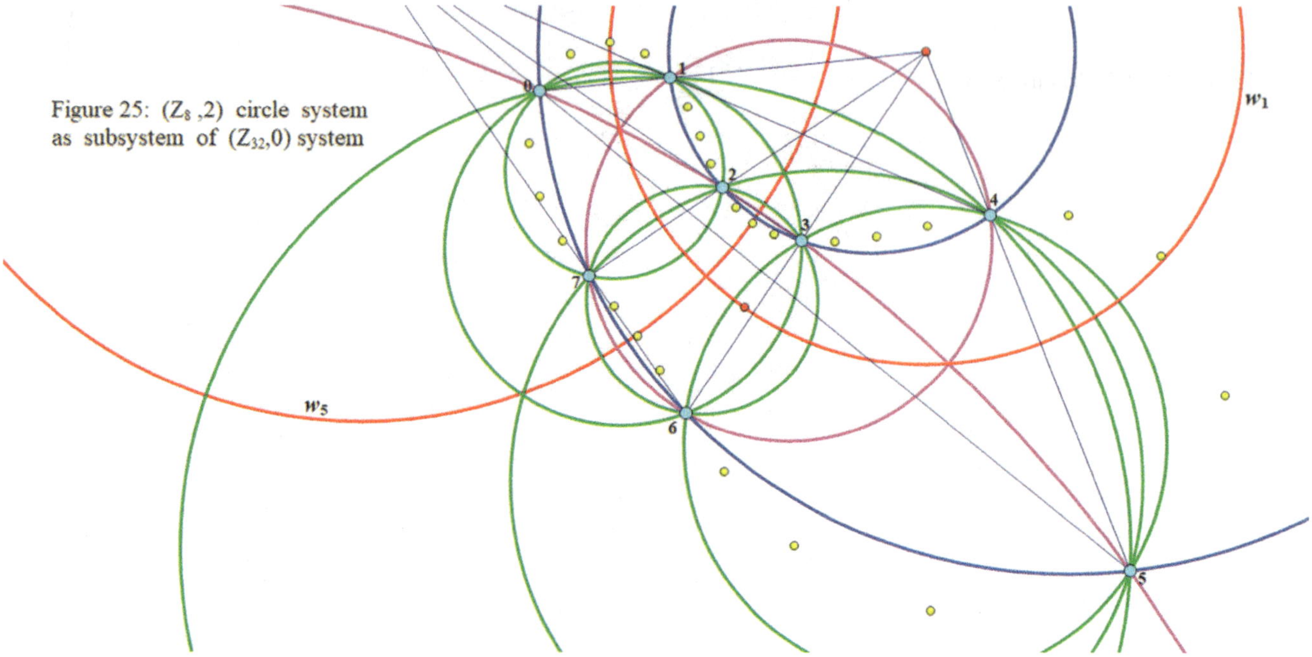

Figure 25: $(Z_8,2)$ circle system as subsystem of $(Z_{32},0)$ system

Theorem 29 *A $(Z_n,0)$ circle system with n=2m can be constructed on any irreducible self-inversive cubic curve γ as follows. (i) Let w_0 be a circle with center X which inverts γ onto itself and which has nonempty intersection with γ as in Figure 25, and put vertex 0 on w_0. (In accordance with Theorem 2 each pair of points {q, -q} will be inverses with respect to w_0. To make the construction more facile, we are supposing that γ is a Type IV cubic and putting vertex 0 on the oval.); (ii) Put vertex -1 on the bell and let vertex 1 denote the inverse of vertex -1 with respect to w_0, thus 1 = X*-1; (iii) If vertices {0, ± 1, ± 2, ..., ± (k-1)} have been defined, let k = X * {(X*0)*(-1 * (-k+1))} = X*{0*(-1*(-k+1))}, and let -k = X*k. Finally, the generating vertex, (vertex -1) is shifted on the bell until points m, -m coincide.*

Proof First we will show that 0*k = -m * (k+m) for all integers m, k. Consider: k = X*{0*(-1*(-k+1))} and thus -k = X*k = 0*(-1*(-k+1)), and so 0*-k = -1*(-k+1)). This holds for every integer k so we can replace -k by k to obtain our desired formula with m = 1. Now assume inductively that 0*k = -m * (k+m) for m = 1,2, …, t-1 and consider: -t * (k+t) = (X*t)*{-1 * (0*(k+t-1))} = (t*-1)*{X*(0*(k+t-1))} = (0*(t-1))* {X*(0*(k+t-1))} = (X*(t-1)) * (k+t-1) = -(t-1) * (k + (t-1)) = 0*k, the last equality by the inductive hypothesis. In case m < 0, then -m>0 and we have 0*k = m*(k-m), or replacing k by -k we have 0*-k = m* (-k-m) and thus (X*X)*(0*-k) = (X*X)* {m* (-k-m)}, {(X*0)*(X*-k)}= {(X*m)*(X*(-k-m))},(0*k)= -m*(m+k)} . We have shown that 0*k = -m * (k+m) for all integers m, k. Replacing k by k+q and m by -q we obtain 0*(k+q) = q*k for all integers k,q. Next we show that -m * (m+k+q) = k*q for all integers m,k,q by induction on m. We have just seen that it holds for m = 0. Now assume inductively that -m * (m+k+q) = k*q for all integers k, q and for m = 0, 1,2, …, t-1, and consider: by the inductive hypothesis we have k*q = -(t-1) * (k+q+t-1) = (X*(t-1))*{0* (-1*(k+q+t)} = ((t-1)*0)*{X*(-1*(k+q+t))} = (-1*t)* {X*(-1*(k+q+t))} = (X*t)*(k+q+t) = -t * (t+k+q).

By induction we have shown that -m * (m+k+q) = k*q for all integers k, q and for m≥0. If m<0 then -m>0 and we have m*(-m+k+q) = k*q. We can replace k by -k and q by -q to obtain m*(-m-k-q) = -k*-q. Then (X*X)*{m* (-m-k-q)} = (X*X)* (-k*-q), (X*m)*(X* (-m-k-q) = (X*-k)*(X*-q), -m*(k+q+m) = k*q as desired. If a,b,c,d ∈ Z and a+b = c+d then a*b = 0*(a+b) = 0*(c+d) = c*d. Now recall from Theorem 16 that points a,b,c,d on γ are cocyclic iff b = X*{(X*a)*(c*d)}. In our case, if a+b+c+d = 0 we have X*{(X*a)*(c*d)} = X*{-a *(c*d)} = X* {(b+c+d)*(c*d)} = (c*-c)* {(b+c+d)*(c*d)} = d*{-c *(c+b+d)} = d*(b*d) = b. Note that in the second to last equality we have used -m * (m+k+q) = k*q with m = c, k = b and q = d. We thus obtain a (Z,0) circle system, and if we shift the generating vertex, -1, so that points m and -m coincide, (they lie together in the oval and on w_0), we obtain a $(Z_{2m},0)$ circle system on γ. □

A $(Z_{32},0)$ circle system is illustrated on a Type IV self-inversive cubic in Figure 26. If m = 2q is even then by using the same procedure as described in the opening paragraph of this Section, we obtain $(Z_q,1)$ and $(Z_q,2)$ subsystems with cubic envelope γ. In case m is odd, i.e., gcd(n,4) = 2, then by Theorem 6 there are two equivalence classes of circle systems, namely those equivalent to $(Z_n,0)$ and those equivalent to $(Z_n,1)$. These we can obtain as subsytems of a $(Z_{4n},0)$ circle system. We saw in Section 3 that a $(Z_8,2)$ circle system exists in which points {p,q} with p+q = 1 are antiinverse pairs. The $(Z_8,2)$ circle system we have constructed in Figure 25 as a subsystem of $(Z_{32},0)$ does not antiinvert onto itself. Thus if gcd(n,4) = 4 it remains to construct a $(Z_n, 2)$ circle system which is antiinversive.

Figure 26: $(Z_{32},0)$ circle system on Type IV cubic curve

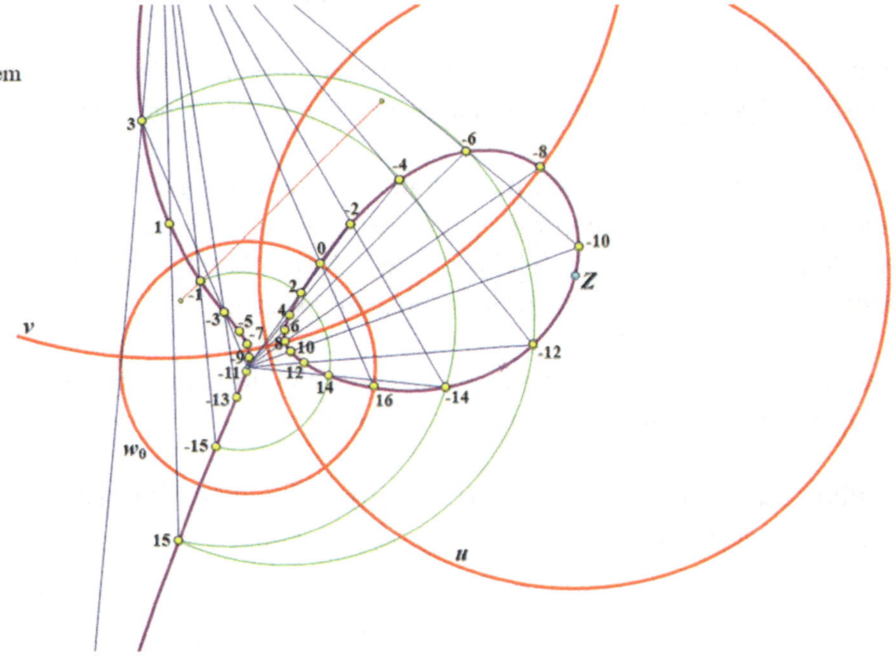

4.2 Perfect polygons and Justified Circle Systems

In Chapter 2 geometric triple systems were used to construct circle systems. Here we introduce another construction method involving *perfect polygons*. In the proof of Theorem 28 we showed that if a+b = c+d then a*b = c*d. This is actually a defining property of a *perfect polygon,* (See Section 1.1). More generally if G is an abelian group then a *perfect G-gon with base G* is defined as follows. Let ψ: G→Π be an injective mapping from G into the projective plane Π such that no three points in ψ(G) are collinear. For each g ∈ G let W_g = {[ψ(a),ψ(b)] : a+b = g}. If for each g ∈ G the lines in W_g are concurrent, then we call the set of points ψ(G) a *perfect G-gon.* If X_g denotes the point of concurrence of the lines in W_g, then the set X = {X_g: g ∈ G} is called the *perspective set* of the perfect G-gon. In case G is the abelian group Z_n then the corresponding perfect G-gon is called a *perfect polygon with base Z_n* or simply a *perfect n-gon* as in Section 1.1. In [1] it is shown that the combined set of vertices and perspective of a perfect n-gon lie on an irreducible cubic curve, and that a perfect n-gon can be constructed on any such curve. The relationship between perfect polygons and circle systems is given in the following results.

Lemma 30 *Let Ω be a $(Z_n,0)$ or $(Z_n,1)$ circle system with a self-inversive cubic envelope λ. If $n \geq 8$ then Ω is a perfect n-gon on λ. If Γ is a $(Z_n,2)$ circle system on λ and $n \geq 9$, then Γ is a perfect n-gon on λ.*

Proof Suppose first that Ω is $(Z_n,0)$. For n = 8, the circles belonging to Ω are listed in Section 1.2. If (a,b,c,d), (a,b,e,f) are two of these circles then c+d = -a-b = e+f and also we have X * {(X*b)*(c*d)}= δ(b,c,d) = a = δ(b,e,f) = X * {(X*b)*(e*f)} when c*d = e*f follows by cancellation. Thus for example, circles c_1, c_2 yield 1*2 = 0*3; circles c_3, c_7 yield 1*2 = 4*7, and circles c_4, c_9 give 0*3 = 5*6. So we obtain the perspective point X_3 = 5*6 = 0*3 =1*2 = 4*7. Via the same considerations, the remaining perspective points are easily obtained, and thus Ω is a perfect octagon. Similar computations show that if Ω is $(Z_n,1)$ then it is also a perfect octagon. If n > 8, the abundance of circles in Ω makes it even easier to prove all perspective points. We shall omit the details. In Figure 27 we illustrate a $(Z_7,0)$ circle system on the oval of a self-inversive Type IV cubic which is neither a perfect 7-gon, nor justified. The $(Z_8,2)$ circle system with 10 circles is not necessarily a perfect octagon as is evidenced by Figure 28. If n >8 the abundance of circles in Γ assures that Γ is a perfect n-gon, again we omit the details. □

Lemma 31 *Let Ω be a (G,g) circle system with a self-inversive cubic envelope λ and suppose Ω is also a perfect G-gon on λ. If for every point k of Ω, there exists a pure circle of Ω which contains k, then Ω is justified on λ.*

Proof To prove justification we must show that δ(a,a,b) = g-2a-b. Let k = g-2a-b, then by hypothesis Ω contains a pure circle (c,d,e,k) which contains k and thus c+d+e + k = g, i.e., c+d+e = 2a+b. Then δ(a,a,b) = X*{(X*b)*(a*a)} = X*{(X*b)*(X_{2a})} = X*{(X*b)*(c*(2a-c))} = X*{(X*c)*(b*(2a-c))} = X*{(X*c)* X_{2a+b-c}} = X*{(X*c)* X_{e+d}} = X*{(X*c)* (e*d)} = δ(c,d,e) = k. □

If n≥6 then all circle systems with base Z_n and which form perfect n-gons on their cubic envelope, satisfy the hypothesis of Lemma 31 and thus are justified.

Corollary 32 If *Ω is a $(Z_n,0)$ or $(Z_n,1)$ circle system with a self-inversive cubic envelope λ, and $n \geq 8$ then Ω is justified on λ. If Γ is a $(Z_n,2)$ circle system on λ, and $n \geq 9$ then Γ is justified on λ.* □

Figure 27: $(Z_7,0)$ circle system on self-inversive cubic which is not a perfect polygon.

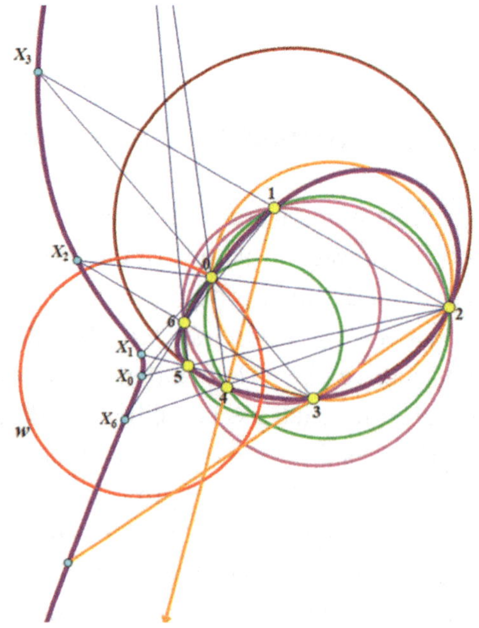

Figure 28: $(Z_8,2)$ circle system on self-inversive cubic which is not a perfect octagon

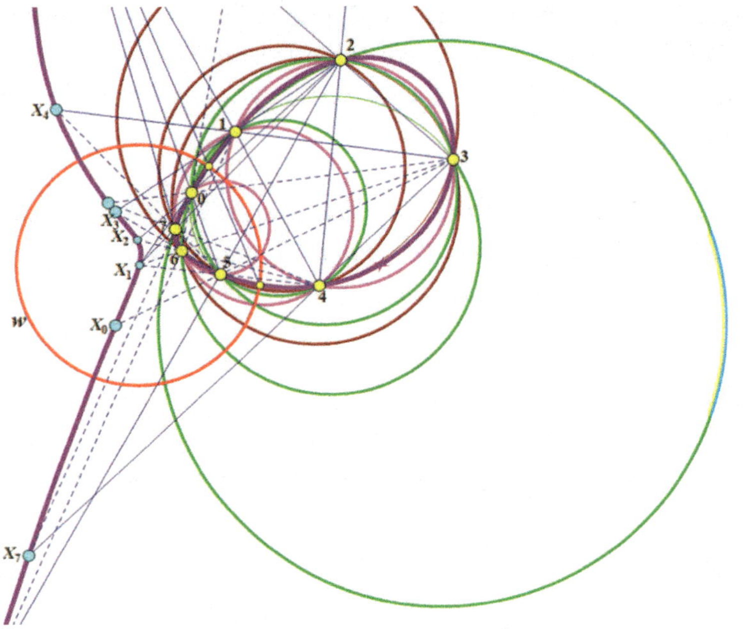

We note that by Theorem 13, if Ω is a $(Z,0)$ or $(Z_n,0)$ circle system and $n \geq 11$, then it has a self-inversive cubic envelope λ, and thus by Corollary 32, Ω must be justified on λ. We also note that clearly the foregoing Lemmas and Corollary apply to any circle system with base Z. A perfect n-gon can be constructed on any nonsingular irreducible cubic λ with point 0 chosen to be any nonflex point of λ. If X is the center of an inverting circle for λ and we choose point 0 so that $X_0 = 0*0 = X$, then the corresponding perfect n-gon is a justified $(Z_n,0)$ circle system in accordance with the following Theorem 34. In Figure 29 we illustrate a $(Z_7,0)$ circle system justified on the oval of a Type IV cubic, obtained by constructing a perfect 7-gon with $X_0 = 0*0 = X$,

Figure 29: justified $(Z_7,0)$ circle system obtained from perfect 7-gon on TypeIV self-inversive cubic curve with X_0 at center of inverting circle w.

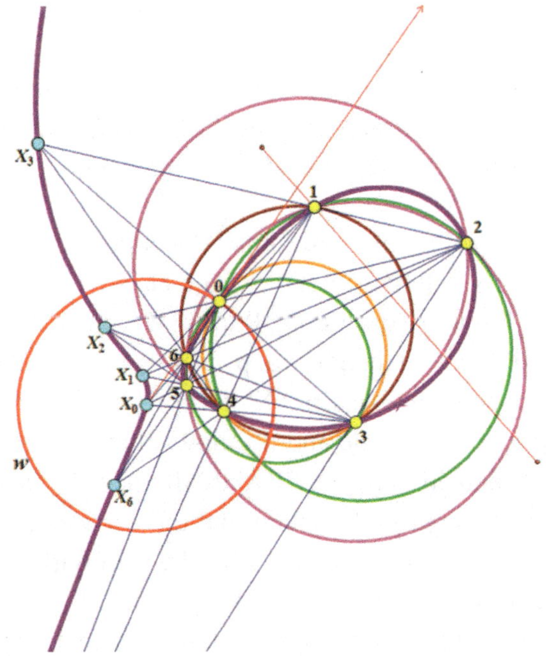

In [1] it is shown that the perspective points of a perfect polygon, if no three are collinear, form a perfect polygon themselves. If $X = \{X_k : k \in Z_n\}$ is the set of perspective points of a perfect n-gon P, then the set of lines $\{[X_i, X_j] : i+j = k\}$ concur at a point W_k and $W = \{W_k : k \in Z_n\}$ is the set of perspective points of the perfect n-gon with vertex set X. These points are indicated for the perfect 7-gon of Figure 29, in Figure 30. If P lies on a self - inversive cubic λ, as in Figure 30, then we have $X_i * X_j = W_{i+j}$ for all i, j in Z_n. If in addition, the vertices of P form a circle system with X_0 the center of an inverter for λ, then we obtain the following circles.

Theorem 33 *Let Ω be a $(Z_n, 0)$ circle system justified on a self-inversive cubic envelope λ. Let $X = \{X_k : k \in Z_n\}$ denote the perspective set of the corresponding perfect n-gon, and let $W = \{W_k : k \in Z_n\}$ denote the perspective set for the perfect n-gon with vertex set X. Then $A = \{(W_p, W_q, X_s, X_t) : s+t = p+q\}$ and $B = \{(W_p, s, t, X_q) : s+t = q-p\}$ are sets of circles.*

Proof To prove the circles in A, consider:

$$\delta(W_p, W_q, X_s) = X_0 * \{(X_0 * W_p) * (W_q * X_s)\} = X_0 * \{X_p * X_{q-s}\} = X_0 * W_{p+q-s} = X_{p+q-s} = X_t,$$

And to prove the circles in B, consider:

$$\delta(W_p, s, t) = X_0 * \{(X_0 * W_p) * (s*t)\} = X_0 * \{X_p * X_{s+t}\} = X_0 * W_{p+s+t} = X_{p+s+t} = X_q. \ \square$$

Some of the circles in sets A, B are indicated in Figure 30. Of special interest are the circles in A since they determine what we shall call a *bipartite circle system.* W and X are perfect n-gons which comprise the partite sets and the circles in A have the form {(p, q, s, t) : p,q ∈ W, s, t ∈ X, and p+q = s+t}. In general if G, G' are identical abelian groups whose elements label points in the plane such that each of the sets {{p, q, s, t} : p,q ∈ G, s, t ∈ G', and p+q = s+t} is cocyclic, we have a *bipartite circle system with base G.* These structures are a subject for future research; it might be asked if the partite sets must be perfect G-gons or if there exists a cubic envelope, etc.

Theorem 34 *Let γ be an irreducible cubic curve which inverts onto itself via a circle w with center X. Let P be a perfect G-gon with all vertices and perspective points on γ, and suppose P contains four cocyclic vertices {a,b,c,d} with a+b+c+d = g for some g∈G. Then the vertices of P form a (G,g) circle system. If $X_0 = X$ then the vertices of P form a justified (G,0) circle system. If Ω is a (G,g) circle system justified on γ then the points of Ω form the vertex set of a perfect G-gon.*

Proof Suppose e, f ∈ G with e+f = c+d. Since P is a perfect G-gon we have e*f = c*d and thus b = δ(a,c,d) = X * {(X*a)*(c*d)} = X * {(X*a)*(e*f)} = δ(a,e,f). We thus obtain that (a,b,e,f) is a circle. Now suppose p,q,r,s are vertices of P with p+q+r+s = g. Putting e = p and f = c+d – p we obtain that (a, b, p, c+d – p) is a circle. From this we obtain that (q, a+b – q, p, c+d – p) is a circle. Since a+b – q + c+d – p = r+s we have r*s = (a+b – q) * (c+d – p) and thus δ(p,s,r) = X*{(X*p)*(r*s)} = X*{(X*p)*((a+b – q) * (c+d – p))} = δ(p, a+b-q, c+d-p) = q. Then (p,q,r,s) is a circle and it follows that the vertex set of P is a (G,g) circle system. In case $X_0 = X$, then $0*0 = X_0$ and $\delta(0,0,0) = X * \{(X*0)*(0*0)\} = X_0 * \{(X_0 * 0)*(0*0)\} = 0$, so we may take a = b= c= d = 0 in the first part of the proof to obtain that the vertices of P form a (G,0) circle system. Now suppose Ω is any (G,g) circle system on γ and Let a,b,c,d,e ∈ G with a+b = c+d, then X*{(X*e)*(a*b)} = δ(a,b,e) = g – (a+b+e) = g - (c+d+e) = δ(c,d,e) = X*{(X*e)*(c*d)}. It follows by cancellation in the BH algebra on γ, that a*b = c*d, and thus Ω forms the vertex set of a perfect G-gon. □

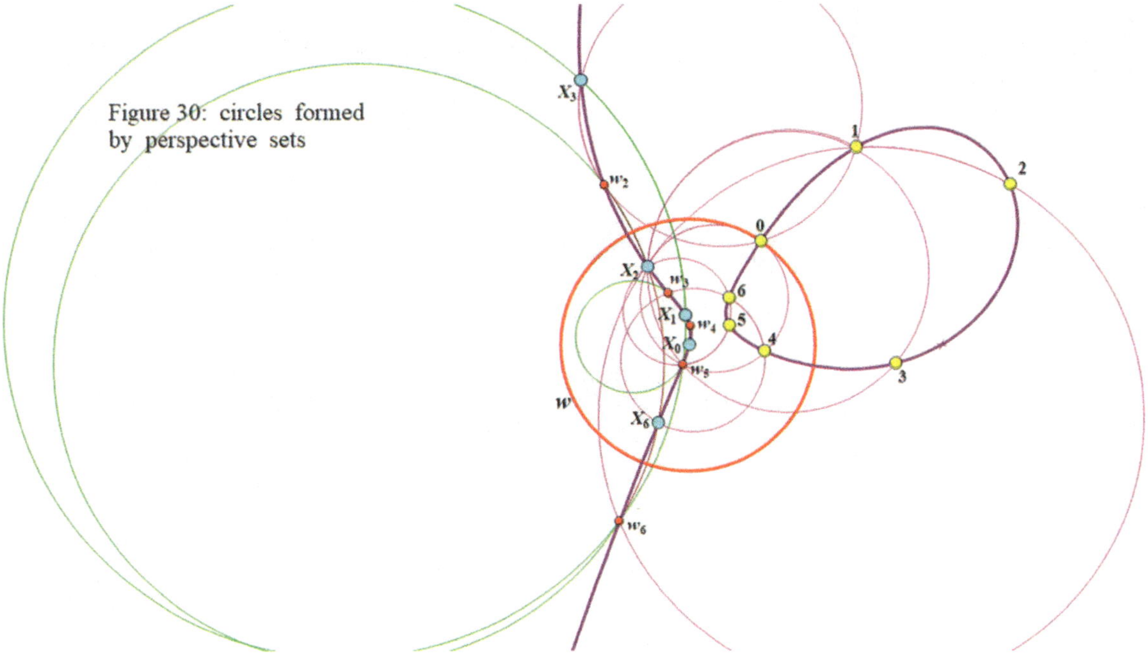

Figure 30: circles formed by perspective sets

In the last paragraph of Section 4.1 it was noted that the construction of an antiinversive $(Z_n,2)$ circle system still needs to be done. Theorem 34 affords us a reasonably facile way to complete this construction.

Theorem 35 *Let γ be a Type IV cubic curve which inverts onto itself via a circle w with center X, and suppose w has nonempty intersection with γ. Suppose n = 2m = 4t is a multiple of 4 and construct a perfect 2n-gon P on the oval of γ with $X_0 = X$. Let S denote the points on the oval of the form $\{4k\text{-}1 : k \in Z_n\}$. If the points of S are relabeled according to $4k\text{-}1 \rightarrow 2k$, we obtain a (E,2) circle system where E denotes the subgroup of Z_n formed by the even values. Let v denote the circle with center Z which inverts the oval onto the bell of γ, and for each $s \in S$ let Z*s have (odd) label (m-s +1) mod n. Then the points in $S \cup (Z*S)$ form a $(Z_n, 2)$ circle system on γ, and all points {p,q} with p+q = 1 are antiinverses of each other. (See Figure 31.)*

Proof Let Ω denote the set of points $S \cup (Z*S)$ with labels as assigned in the Theorem statement, so the even points, (S) lie on the oval and the odd points (Z*S) on the bell. Since the points on the bell form an (E,2) circle system, we have every circle consisting of four even points. The odd points consist of inverses of the even points with respect to circle v, and since inversion preserves circles we also have all circles consisting of four odd points. The only remaining circles of a $(Z_n,2)$ circle system have the form (p,q,u,t) with p+q+u+t = 2 (mod n) where p,q are even and u,t are odd. Let (γ,δ) denote the THA defined on γ. By Theorem 16, the fourth point on a circle (a,b,c) and on the self-inversive cubic γ is given by $\delta(a,b,c) = Z*\{(Z*a)*(b*c)\}$ since Z is the center of a circle (v) which inverts γ onto itself. Thus $\delta(p,q,s) = Z*\{(Z*u)*(p*q)\} = Z*\{(m+1-u) * (p*q)\}$. Now the even points on the oval derive from the perfect 2n-gon P and satisfy a*b = c*d if a+b = c+d. Since m+1-u, p,

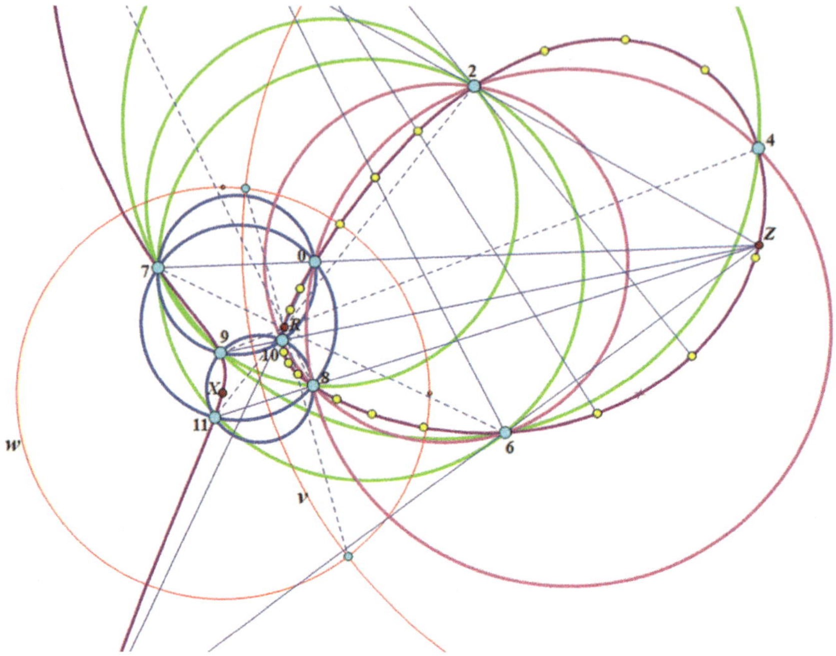

Figure 31: antiinversive $(Z_{12},2)$ circle system

q are all even we have p*q = (m+1-u)*{(p+q) – (m+1-u)}, and thus (m+1-u) * (p*q) = (p+q) – (m+1-u). Then we have δ(p,q,s) = Z*{(p+q) – (m+1-u)} = (m+1) – {(p+q) – (m+1-u)} = (2m+2) –p – q - u = 2 – p – q – u = t. Thus (p,q,u,t) is indeed a circle, and Ω is a $(Z_n,2)$ circle system. According to the discussion preceding Theorem 2 there exist two circles w_1 and w_{m+1} which either invert or antiinvert Ω onto itself. Circle w_1 inverts or antiinverts each pair of points {p,q} with p+q = 1, and circle w_{m+1} either inverts or antiinverts each pair of points {s,t} with s+t = m+1. In our construction we have v = w_{m+1} which inverts Ω onto itself. In [7] and [5] it is shown that every Type IV self-inversive cubic inverts onto itself via three mutually orthogonal circles and antiinverts onto itself via a fourth circle. These four circles each have as their centers the four points on the cubic which satisfy x*x = ∞, where ∞ denotes the unique point at infinity on the cubic. Two of these centers lie on the bell and the other two on the oval. The circles with centers on the bell are inverters. As for the circles with centers on the oval, one is an inverter and the other is an antiinverter. By our construction circle v inverts all pairs {p,q} with p+q = m+1, and so circle v must be an inverter for the entire cubic γ. The center R of circle w_1 must lie on the oval since inverse pairs with respect to w_1 consist of an odd and an even point. It must then be that w_1 antiinverts each pair of points {s,t} with s+t = 1. Then it follows that w_1 antiinverts the entire cubic γ onto itself. □

The argument given at the end of Section 3.2 showing that an inversive (Z,2) circle system Ω has an envelope ξ which inverts onto itself via the circle w_1 which inverts all pairs {p,q} with p+q = 1, can also be used, provided n ≥10, to show that an inversive $(Z_n,2)$, (n = 2m), circle system Γ also has an envelope μ which inverts onto itself via the circle which inverts pairs with sum 1. By Theorem 14 there also exists a second circle v orthogonal to w_1 which also inverts μ onto itself. For the $(Z_n,2)$ system Γ we have v = w_{m+1} the circle which inverts all pairs {s,t} of Ω with s+t = m+1. In case Ω is an antiinversive $(Z_n,2)$ circle system with envelope ξ, then one of w_1, w_{m+1} inverts ξ onto itself, and the other antiinverts ξ onto itself. By Theorem 26 there exist two additional circles which invert ξ onto itself, forming a trio of three mutually orthogonal circles which invert ξ onto itself. In case of a $(Z_n,1)$ circle system Λ it should be possible to double the value on each vertex and then insert odd vertices to obtain a $(Z_{2n},2)$ circle system Λ́. We could then say that the envelope of Λ must be self-inversive, but it is not clear how this construction can be accomplished.

Chapter 5

Circle Systems with Base $Z_2{\times}Z_4$

5.1 The $(Z_2{\times}Z_{4,}(0,0))$ Circle System

Circle systems with base $Z_2{\times}Z_n$ with n even, (n = 2m), form a second major class of circle systems. We can suppose n is even since otherwise the group $Z_2{\times}Z_n$ is isomorphic to Z_{2n} yielding circle systems already considered. In order to introduce the subject of circle systems with base Z_n by simple examples, systems with base Z_8 were considered closely in Chapter 1. We follow the same procedure here by first studying circle systems with base $Z_2{\times}Z_4$. Let Ω be a $(Z_2{\times}Z_4, (0,0))$ circle system. Recall that a circle set is *balanced* in case it consists of two pairs of points with equal sums. As it turns out every circle of Ω is balanced; $\Omega = \Omega_{(1,0)} \cup \Omega_{(1,2)}$ where $\Omega_{(1,0)}$ consists of the circles

((0,0),(1,0),(1,1),(0,3))

((0,0),(1,0),(1,3),(0,1))

((0,0),(1,0),(1,2),(0,2))

((0,2),(1,2),(1,3),(0,1))

((0,2),(1,2),(0,2),(1,1))

((0,1),(1,3),(0,3),(1,1))

and $\Omega_{(1,2)}$ consists of the circles

((0,0),(1,2),(0,1),(1,1))

((0,0),(1,2),(0,3),(1,3))

((1,0),(0,2),(0,1),(1,1))

((1,0),(0,2),(0,3),(1,3)).

We will now give two constructions of $(Z_2{\times}Z_4, (0,0))$ circle system. In accordance with the discussion preceding Theorem 2, there exists a circle $w_{(1,0)}$ which either inverts or antiinverts each pair of points (p,q) with p+q = (1,0), and there exists a circle $w_{(1,2)}$ which either inverts or antiinverts each pair {s,t} with s+t = (1,2). First suppose both $w_{(1,0)}$ and $w_{(1,2)}$ are inverters. To construct Ω we start with three mutually intersecting circles with external radical center S and label the contact points as in Figure 32. Construct circle $w_{(1,0)}$ with center S and orthogonal to circle ((0,0),(1,0),(1,2),(0,2)). Let R = $[(0,0),(1,2)]\cap[(1,0),(0,2)]$ and construct circle $w_{(1,2)}$ with center R orthogonal to circle ((0,0),(1,0),(1,2),(0,2)). Let point (1,1) be the inverse of (0,1) with respect to $w_{(1,2)}$ and let (0,3) be the inverse of (1,3) with respect to $w_{(1,2)}$. By inversion property P2 we obtain immediately all circles in $\Omega_{(1,2)}$. To obtain the circles in $\Omega_{(1,0)}$ we need only show that the points {(0,3),(1,1)} are an inverse pair with respect to circle $w_{(1,0)}$. By inversion property P1, circle α = ((0,2),(1,2),(0,3),(1,1)) is orthogonal to $w_{(1,0)}$ since {(0,2),(1,2)} is an inverse pair with respect to $w_{(1,0)}$. Thus the inverse of (0,3) with respect to $w_{(1,0)}$ must lie on α. Similarly circle β = ((0,0),(1,0),(0,3),(1,1)) is orthogonal to $w_{(1,0)}$ with the consequence that the inverse of (0,3) with respect to $w_{(1,0)}$ must also lie on β. Then the inverse of (0,3) with respect to $w_{(1,0)}$ lies in $\alpha\cap\beta$ = {(0,3), (1,1)} and thus {(0,3),(1,1)} is an inverse pair with respect to $w_{(1,0)}$. □

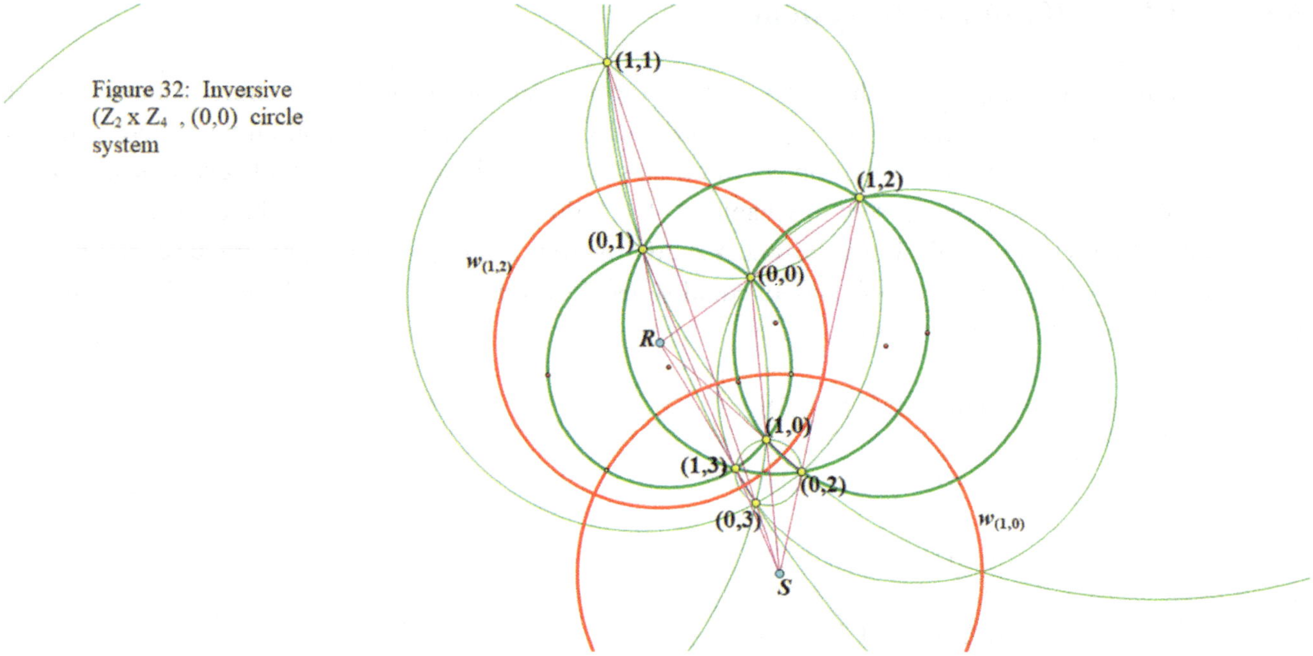

Figure 32: Inversive ($Z_2 \times Z_4$, (0,0) circle system

Note that no matter how the points {(0,0), (0,2), (1,0), (0,0)} are ordered around the corresponding circle of Ω, at least one of the points R, S must lie in the exterior of this circle. Consequently at least one of the circles $w_{(1,0)}$, $w_{(1,2)}$ must be an inverter.

To construct an antiinversive ($Z_2 \times Z_4$,(0,0)) circle system Γ we start with three mutually intersecting circles whose radical center R lies in the interior of the circles, and then label the contact points as in Figure 33. Let $w_{(1,0)}$ denote the circle with center R which antiinverts each of the three circles onto themselves. The remaining two points of Γ are defined by letting (0,3) equal the intersection point, besides (1,3) of circles ((1,3),(0,2),(1,0)), ((1,3),(0,0),(1,2)), and letting (1,1) denote the contact point, besides (0,1) of circles ((0,1),(0,0),(1,2)), ((0,1),(0,2),(1,0)). To prove the ten circles of Γ, it suffices to show that {(0,3),(1,1)} is an antiinverse pair with respect to $w_{(1,0)}$. By antiinverse property P8, The antiinverse of circle ((0,3),(0,0),(1,2),(1,3)) is circle α = ((1,0),(0,2),(0,1)), and the antiinverse of circle ((0,3),(0,2),(1,3),(1,0)) is circle β = ((1,2),(0,1),(0,0)). So the antiinverse of (0,3) must lie in $\alpha\cap\beta$ = {(0,1),(1,1)}. Since (0,1) is the antiinverse of (1,3) with respect to $w_{(1,0)}$ we can conclude that {(0,3), (1,1)} is an antiinverse pair with respect to $w_{(1,0)}$.

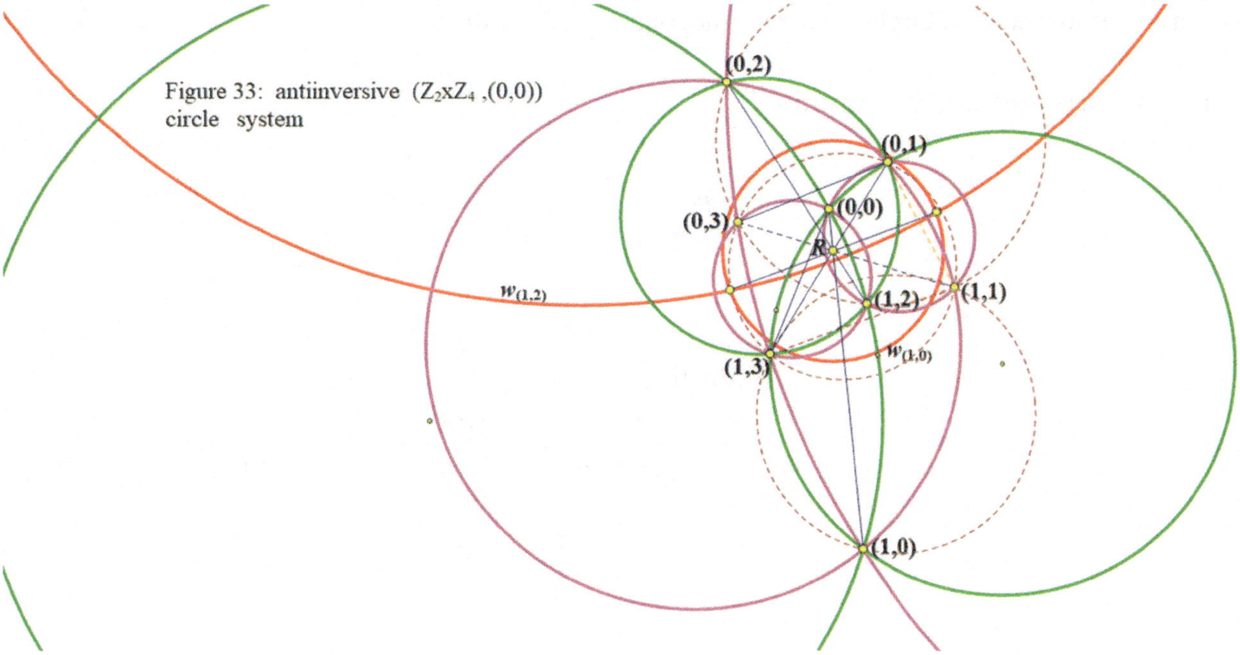

Figure 33: antiinversive (Z_2xZ_4 ,(0,0)) circle system

5.2 The ($Z_2 \times Z_4$,(0,2)) circle system

Next we construct a ($Z_2 \times Z_4$,(0,2)) circle system Ω. The solutions to s+s = (0,2) are given by {(0,1),(1,1),(0,3),(1,3)}. These have respective corresponding circle complexes $\{\Omega_{(0,1)}, \Omega_{(1,1)}, \Omega_{(0,3)}, \Omega_{(1,3)}\}$ and inverting (or antiinverting) circles $w_{(0,1)}$, $w_{(1,1)}$, $w_{(0,3)}$, $w_{(1,3)}$. We will show that three of these are mutually orthogonal inverters and the remaining circle is an antiinverter. To accomplish this we will make use of the following Lemma which is the converse of Lemma 1.

Lemma 36 *Let u,v be circles and ϕ_u, ϕ_v inversion mappings via u, v respectively. If there exists a point P not on u or v such that $\phi_u\,\phi_v(P) = \phi_v\,\phi_u(P)$ then circles u, v are orthogonal. Moreover, if J,K denote the contact points of u, v then the points P,J, $\phi_v\,\phi_u(P)$, K are cocyclic.*

Proof We will set up an analytic proof by letting u be the unit circle and v be the circle with center (a,0) and radius r, as in Figure 34. Let P have coordinates (s,t) and let M = $s^2 + t^2$; Q = $(s-a)^2 + t^2$, and K = $(t/M)^2 + ((s/M)-a)^2 = \frac{1-2as+Ma^2}{M}$. Then $\phi_v(P) = (\frac{aQ+(s-a)r^2}{Q}, \frac{tr^2}{Q})$; $\phi_u(P)$ = (s/M, t/M) and $\phi_v\,\phi_u(P) = (\frac{ak+(\frac{?}{M})-a)r^2}{K}; \frac{(\frac{t}{M})r^2}{K})$. Now the condition $\phi_u\,\phi_v(P) = \phi_v\,\phi_u(P)$ will be satisfied if points $\phi_v(P)$, $\phi_v\,\phi_u(P)$ lie on a line through the origin. This occurs if

$$\frac{tr^2}{aQ+(s-a)r^2} = \frac{tr^2}{aMK+(s-aM)r^2},$$

$$aQ + (s-a)r^2 = aMK + (s-aM)r^2,$$

$$r2 = \frac{Q-MK}{1-M} = \frac{t2+(s-a)^2-1+2as-Ma^2}{1-M} = \frac{M+a^2-1-Ma^2}{1-M} = \frac{(M-1)(1-a^2)}{1-M}\ a^2\text{-}1$$

From $r^2 = a^2 - 1$ it follows that circles u,v are orthogonal. To prove the second

Figure 34: orthogonality Lemma

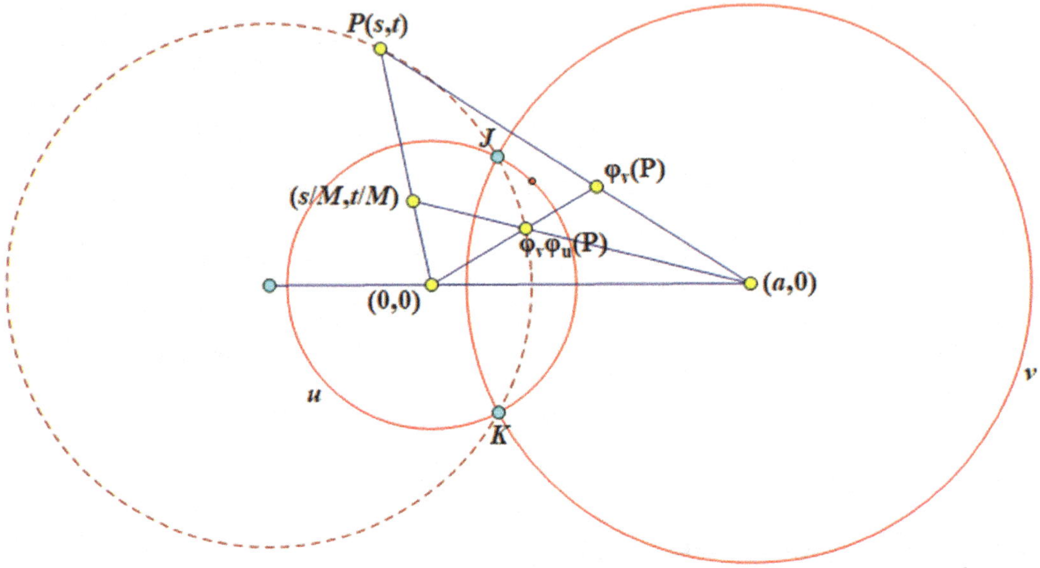

statement of the Lemma, invert Figure 34 by a circle with center J. If x is a point, let x' denote the inverse of x with respect to this circle. Then u', v' are perpendicular lines which meet at K', and the points P', $(\phi_v(P))'$, $(\phi_v\,\phi_u(P))'$, $(\phi_u(P))'$ form a rectangle which is symmetric with respect to both of the lines u', v'. The line [P', $(\phi_v\,\phi_u(P))'$] is a diagonal of this rectangle, and so passes through the center of the rectangle K'. Thus points {P', K', $(\phi_v\,\phi_u(P))'$} are collinear with the consequence that points {J,K, P, $\phi_v\,\phi_u(P)$)} are cocyclic. □

Now if all four of the circles $w_{(0,1)}$, $w_{(1,1)}$, $w_{(0,3)}$, $w_{(1,3)}$ are inverters, then by Lemma 36 we would have four mutually orthogonal circles, which is not possible. We conclude that at most three of these circles are inverters and at least one must be an antiinverter. To show that in fact there is exactly one antiinverter consider the circles c_1= ((0,0),(1,1),(0,1),(1,0)), c_2 = ((0,0),(1,3),(0,1),(1,2)), c_3 = ((0,0),(0,3),(0,1),(0,2)), c_4 = ((0,0),(0,3),(1,1),(1,2), c_5 = ((0,0),(1,3),(1,1),(0,2)) and c_6 = ((0,3),(1.3),(0,0),(1,0) where the points occur in clockwise order as listed. Circles c_1, c_2, c_3 show that if $w_{(0,1)}$ is an antiinverter then the other three circles $w_{(1,1)}$, $w_{(0,3)}$, $w_{(1,3)}$ must be inverters. Circles c_4, c_5, c_6 show similarly that no two of the circles $w_{(1,1)}$, $w_{(0,3)}$, $w_{(1,3)}$ can be antiinverters. With the same considerations the following general result can be proved.

Theorem 37 *Let Ω be a ($Z_2 \times Z_{2m}$, (0,2)) circle system and let $w_{(0,1)}$, $w_{(1,1)}$, $w_{(0,m+1)}$, $w_{(1,m+1)}$ denote the four circles which invert (or antiinvert) Ω onto itself. Then exactly one, (any one), of these circles is an antiinverter and the other three are mutually orthogonal inverters.*

Proof It remains only to show that any one of the four circles $w_{(0,1)}$, $w_{(1,1)}$, $w_{(0,m+1)}$, $w_{(1,m+1)}$ can serve as the antiinverter. Suppose $w_{(0,1)}$ is the antiinverter and m is even, then the relabeling of Ω given by (s,t)→(s,t) if t is even and (s,t)→(s, t+m) if t is odd is an equivalence which yields a ($Z_2 \times Z_{2m}$, (0,2)) circle system in which $w_{(0,m+1)}$ is the antiinverter. The same relabeling converts a $w_{(1,1)}$ antiinverter into a $w_{(1,m+1)}$ antiinverter. If m is odd, and $w_{(0,1)}$ is an antiinverter, then the relabeling: (0,t)→(1, t+m) if t is odd; (0,t)→(0,t) if t is even; (1,t)→(1, t+m) if t is even, and (1,t)→(1,t) if t is odd, results in a ($Z_2 \times Z_{2m}$, (0,2)) circle system in which $w_{(1,m+1)}$ is an antiinverter. Thus by relabelings we can make any one of the four circles $w_{(0,1)}$, $w_{(1,1)}$, $w_{(0,m+1)}$, $w_{(1,m+1)}$ the antiinverter. □

Now for our construction of the $(Z_2 \times Z_4, (0,2))$ circle system Ω we may start with three mutually orthogonal circles in accordance with Theorem 37. We will select $w_{(0,1)}$, $w_{(1,1)}$, and $w_{(0,3)}$ as our mutually orthogonal inverters with radical center R, and circle $w_{(1,3)}$ as our antiinverter as in Figure 34. The construction then proceeds as follows. (i) let (0,0), (0,3) be inverses with respect to $w_{(0,3)}$, and let (1,1), (1,2) be inverses of (0,0), (0,3) respectively with respect to $w_{(1,1)}$. By Lemma 1, (1,1), (1,2) are inverses with respect to $w_{(0,3)}$; (ii) let (1,0),(1,3) be the inverses of (1,1),(1,2) respectively with respect to $w_{(0,1)}$. By Lemma 1, (1,0), (1,3) are inverses with respect to $w_{(0,3)}$; (iii) Let (0,1), (0,2) be the inverses of (0,0),(0,3) respectively with respect to $w_{(0,1)}$. Then by Lemma 1, (0,1), (0,2) are inverses with respect to $w_{(0,3)}$, and also (1,3), (0,2) are inverses with respect to $w_{(1,1)}$. In short, each of the circles $w_{(0,1)}$, $w_{(1,1)}$, and $w_{(0,3)}$ inverts respectively all pairs of points with sums (0,1), (1,1), (0,3). Each of the 12 circles of Ω is balanced and we have proved, as a consequence of inversion property P2, that every circle in the complexes $\Omega_{(0,1)}$, $\Omega_{(1,1)}$, $\Omega_{(0,3)}$ is obtained by our construction. It remains to prove the circles in $\Omega_{(1,3)}$. To do this it suffices by antiinverse property P7 that each pair of points (p,q) with p+q = (1,3) are antiinverses with respect to the circle $w_{(1,3)}$ with center R which antiinverts each of $w_{(0,1)}$,$w_{(1,1)}$ and $w_{(0,3)}$ onto themselves. To accomplish this we use the following Lemma.

Lemma 38 *Let c_1, c_2, c_3 be three mutually orthogonal circles and let ϕ_1, ϕ_2, ϕ_3 be inversion mapings through c_1, c_2, c_3 respectively. If X is any point in the plane not at the center of c_1, c_2, or c_3 then the points {x, $\phi_1\phi_2(X)$, $\phi_3(X)$, $\phi_1\phi_2\phi_3(X)$, J, K} are cocyclic, where J, K are the points of contact of c_1, c_2. If R is the radical center of c_1, c_2, c_3 then there exists a circle v with center R such that the antiinverse of X with respect to v is the point $\phi_1\phi_2\phi_3(X)$. (See Figure 36.)*

Proof Let M denote the contact point of c_2, c_3 which lies in the interior of c_1 and invert Figure 36 by a circle w with center M. The result is illustrated in Figure 37. Since c_2, c_3 are orthogonal they invert into perpendicular lines which we may take to be the y, x-axes respectively. Since c_1 is orthogonal to both c_2, c_3 it must invert to a circle c_1' with center at the origin. We may take c_1' to have unit radius when the points J', K' will have coordinates (0,1) and (0,-1) respectively as in Figure 37. Let X' = (p,q), then $\phi_3(X)'$ = (p,-q) and $\phi_2(X)'$ = (-p,q). The inverse of $\phi_2(X)'$ with respect to the unit circle is $\phi_1\phi_2(X)' = (\frac{-p}{p^2+q^2}, \frac{q}{p^2+q^2})$ and $\phi_1\phi_2\phi_3(X)'$ is the reflection of $\phi_1\phi_2(X)'$ across the x-axis, so $\phi_1\phi_2\phi_3(X)' = (\frac{-p}{p^2+q^2}, \frac{-q}{p^2+q^2})$. Circle c_4 = (J',K',X') has an equation of the form $(x-h)^2 + y^2 = r^2$. Substituting the coordinates of J' and X' into this equation, we obtain $h = \frac{p^2+q^2-1}{2p}$ and $r^2 = h^2 + 1$. The equation of c_4 then becomes $x^2 - (\frac{p^2+q^2-1}{p})x + y^2 = 1$. The coordinates of $\phi_1\phi_2(X)'$, $\phi_3(X)'$, $\phi_1\phi_2\phi_3(X)'$ are easily seen to satisfy this equation. As a consequence the first part of the Lemma is proved. Now let M,N denote the contact points of c_2, c_3 and let G,H denote the contact points of c_1, c_3. Then we obtain similarly the cocyclic sets c_5 = (M, N, X, $\phi_1(X)$, $\phi_2\phi_3(X)$, $\phi_1\phi_2\phi_3(X)$), c_6 = (G, H, X, $\phi_2(X)$, $\phi_1\phi_3(X)$, $\phi_2\phi_3(X)$). Let v be the circle with center R which antiinverts c_1, c_2, c_3 onto themselves, then points J,K are antiinverses with respect to v. By antiinverse property P9, c_4 antiinverts onto itself via circle v. Similarly, circles c_5, c_6 antiinvert onto themselves via circle v. Since X lies on c_4 and c_5, the antiinverse of X with respect to v lies in $c_4 \cap c_5$,i.e., $\phi_1\phi_2\phi_3(X)$ is the antiinverse of X with respect to v. □

Figure 35: ($Z_2 \times Z_4$, (0,2)) circle system

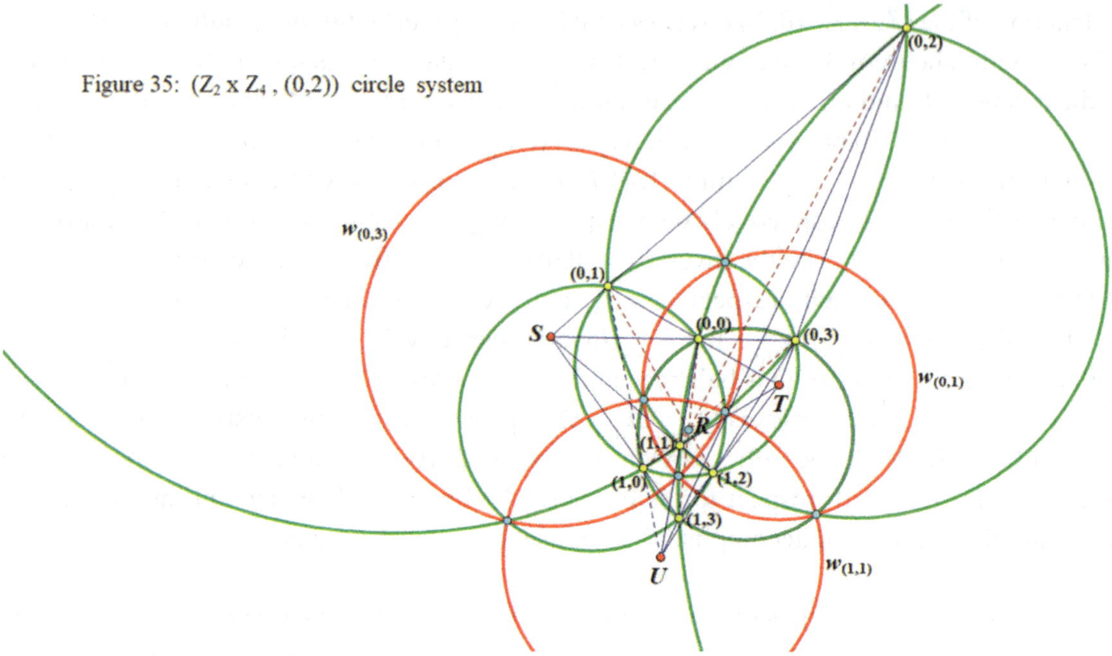

Figure 36: 6-point circles

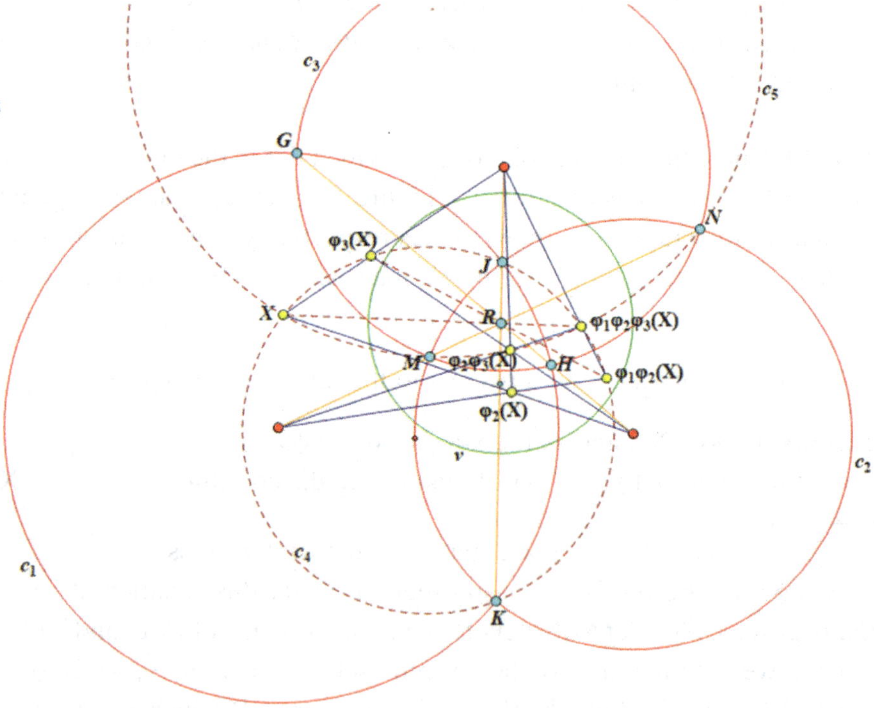

Figure 37: inverse of Figure 31

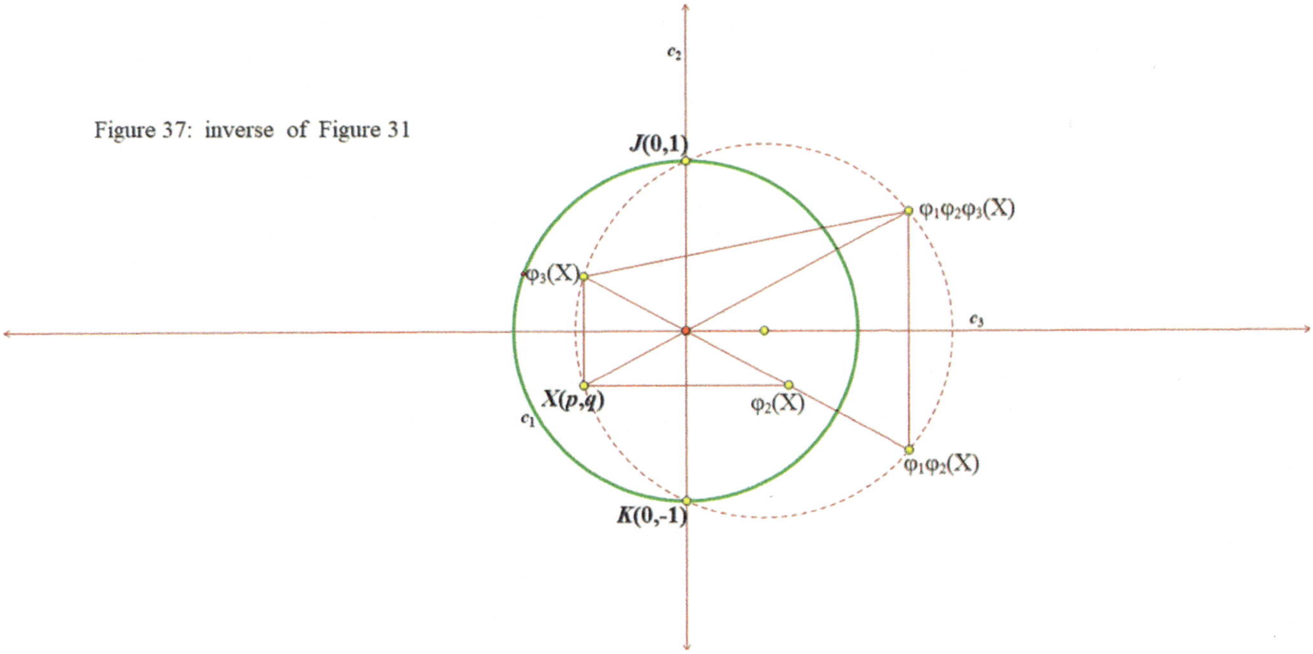

Now to complete the proof of our construction of the $(Z_2 \times Z_4, (0,2))$ circle system Ω, let ϕ_v denote the antiinversion mapping through v, and let $\phi_{(0,1)}, \phi_{(1,1)}, \phi_{(0,3)}$ denote inversions through circles $w_{(0,1)}, w_{(1,1)}$ and $w_{(0,3)}$ respectively. By Lemma 38, $\phi_v(s,t) = \phi_{(0,1)}\phi_{(1,1)}\phi_{(0,3)}(s,t) = \phi_{(0,1)}\phi_{(1,1)}(s, 3\text{-}t) = \phi_{(0,1)}(s+1, t\text{-}2) = (s+1, 3\text{-}t)$. The sum of (s,t) and (s+1, 3-t) in $Z_2 \times Z_4$ is (1,3), so we see that every pair of points with sum (1,3) is an antiinverse pair. We then obtain all circles in the $\Omega_{(1,3)}$ complex thus completing the proof.

5.3 The $(Z_2 \times Z_4,(1,0))$ and $(Z_2 \times Z_4,(0,1))$ circle systems

In Figure 38 we have constructed a $(Z_2 \times Z_4,(1,0))$ circle system as follows: (i) put four points {(1,3),(0,2),(1,1),(1,2)} on a circle as in the Figure; (ii) put points (0,0), (0,3) on a circle through (0,2), (1,3); (iii) Let point (0,1) be the intersection of circles ((0,3),(0,2),(1,2)) and ((0,0),(0,2),(1,1)) besides (0,2); (iv) Let (1,0) be the point of intersection of circles ((1,2),(1,1),(0,1)) and ((1,3),(1,1),(0,0)) besides (1,1). It remains only to prove circles α = ((1,3),(0,3),(1,0),(1,2)) and β = ((0,0),(0,1),(1,0),(0,3)). Consider the circle configurations:

(0,0)   (0,2)   (1,1)   (0,2)
(1,3)   (0,3)   (0,0)   (0,1)
(1,0)   (1,2)   (1,0)   (0,3)
(1,1)   (0,1)   (1,3)   (1,2)

Figure 38: (Z_2xZ_4 , (1,0)) circle system

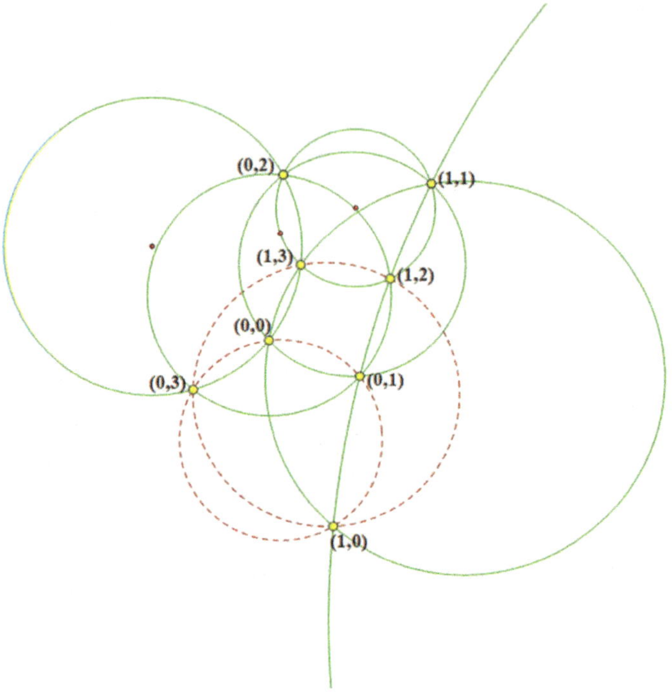

The left configuration proves α, and then the right configuration proves β. The ($Z_2 \times Z_4$,(0,1)) system is illustrated in Figure 39. The construction is accomplished as follows.

Figure 39: (Z_2xZ_4,(0,1) circle system

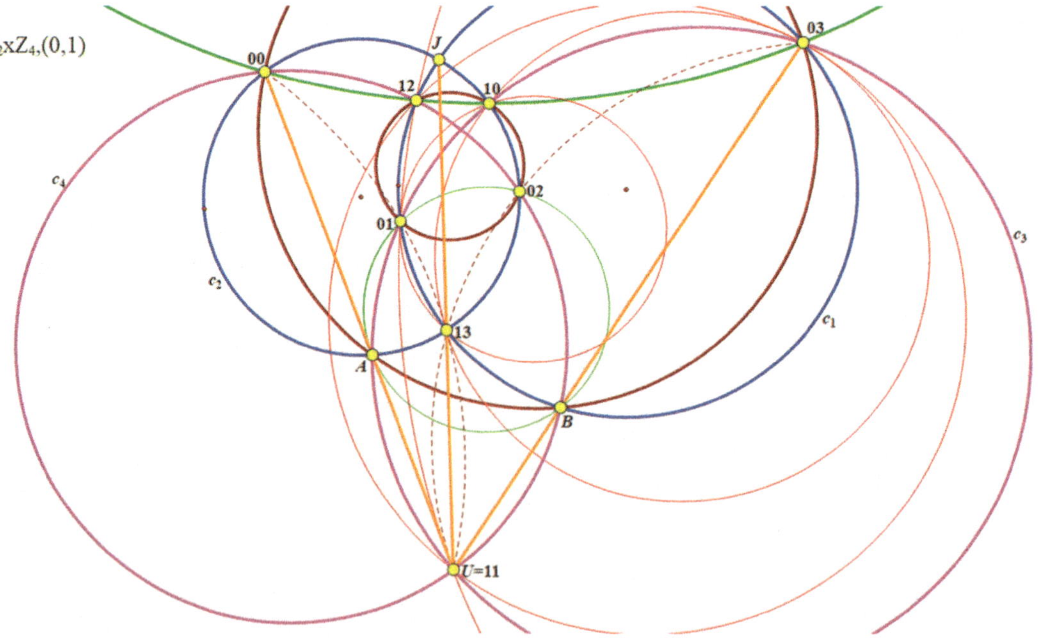

Theorem 39 *The following construction yields a ($Z_2 \times Z_4$, (0,1)) circle system. (i) Let circles c_1, c_2 meet at points J, (1,3) as in Figure 39; (ii) Put point (1,1) on line [J, (1,3)] in the exterior of circles c_1, c_2 ; (iii) Put (0,0) on c_2 and (0,3) on c_1 ; (iv) Let A denote the contact point of c_2 and line [(0,0),(1,1)] different from (0,0), and let B denote the contact point of c_1 and line [(0,3),(1,1)] different from (0,3); (v) let c_3 denote circle (A, (1,1), (0,3)) and let c_4 denote circle (B, (1,1),(0,0)) as in Figure 39; (vi) shift point (0,3) on c_3 until {(0,0),(1,2),(1,0),(0,3)} is a cocylic set.*

Proof Construct a circle w with center (1,1) which is orthogonal to c_1. This circle is also orthogonal to c_2 and thus the pairs {(0,0),A} and {(0,3), B} are inverse pairs with respect to w. By inverse property P2, ((0,0), A, (0,3), B) is a circle. The circle configuration:

(0,3)			A
	(0,1)	(1,0)	
	(1,2)	(0,2)	
B			(0,0)

then shows that ((0,1),(1,0),(1,2),(0,2)) is a circle. This circle is maintained when we shift (0,3) on c_1 until circle ((0,0),(1,2),(1,0),(0,3)) is obtained. The circle configuration:

(1,1)			(1,2)
	(0,3)	(0,3)	
	(1,3)	(1,0)	
(0,2)			(0,0)

Proves the tangency of the auxiliary circles ((1,1),(1,2),(0,3)), ((1,0),(1,3),(0,3)) at (0,3). Then the circle configuration:

(1,0)			(0,0)
	(1,3)	(0,2)	
	(0,3)	(1,1)	
(0,3)			(1,2)

Proves circle ((1,3),(0,2),(0,3),(1,1). The circle configuration:

(1,0)			(1,3)
	(0,1)	(0,1)	
	(1,1)	(1,2)	
(0,3)			(0,3)

uses the above proved tangency at (0,3) to prove the tangency of auxiliary circles ((1,0),(1,3),(0,1)), ((1,1),(1,2),(0,1)) at (0,1). Finally the circle configuration:

(0,2)			(1,0)
	(0,0)	(1,3)	
	(1,1)	(0,1)	
(1,2)			(0,1)

proves circle ((0,0),(1,3),(1,1),(0,1)). The eight circles of a ($Z_2 \times Z_4$, (0,1)) circle system have now been proved. □

5.4 The $(Z_2 \times Z_4, (1,1))$ circle system

The last circle system with base $Z_2 \times Z_4$ is the $(Z_2 \times Z_4, (1,1))$ circle system illustrated in Figure 40. The following is the construction and proof.

Figure 40: $(Z_2xZ_4, (1,1))$ circle system

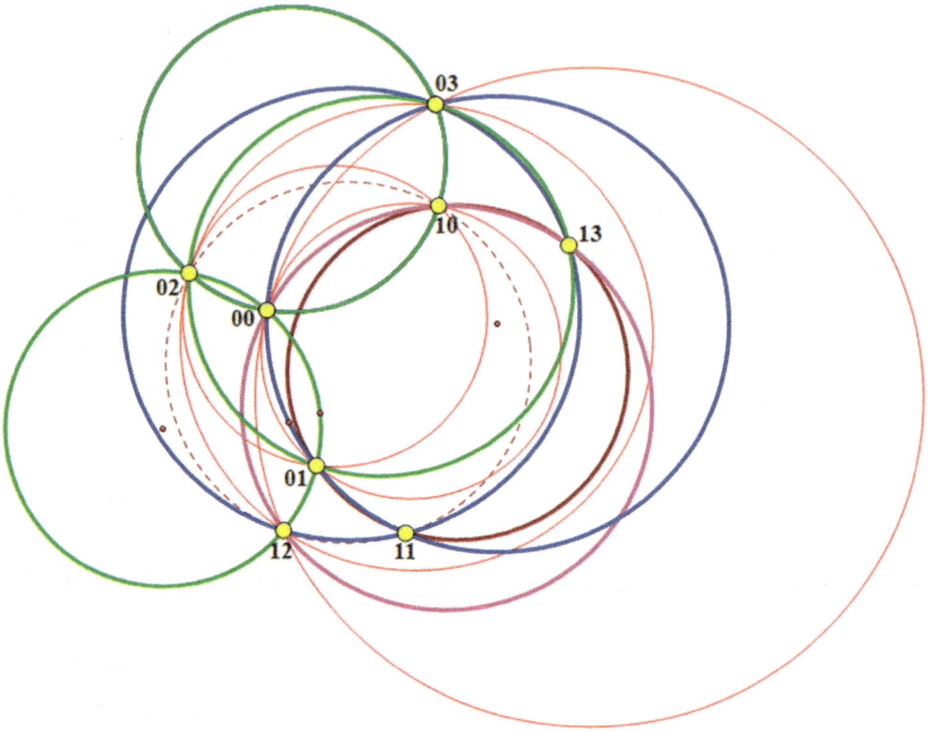

Theorem 40 *Some point shifting is required to obtain the seven bold circles in Figure 40. These constitute all but one of the circles in a* $(Z_2 \times Z_4, (1,1))$ *circle system . The remaining (dashed) circle ((1,0),(1,1),(1,2), (0,2)) is then obtained.*

Proof The circle configuration:

(0,1)			(1,0)
	(0,0)	(0,0)	
	(0,3)	(1,2)	
(1,1)			(1.3)

Proves the tangency at (0,0) of the auxiliary circles ((0,0),(0,1),(1,0)) and ((0,0),(0,3),(1,2)). Then the circle configuration:

(1,2)			(0,3)
	(0,2)	(0,2)	
	(0,1)	(1,0)	
(0,0)			(0,0)

Proves the tangency at (0,2) of the auxiliary circles ((0,2),(1,2),(0,3)) and ((0,2),(0,1),(1,0)), and then the circle configuration:

(1,3)			(0,3)
	(1,1)	(1,2)	
	(1,0)	(0,2)	
(0,1)			(0,2)

Proves the desired circle ((1,0),(1,1),(1,2), (0,2)). □

5.5 Parameter Comparisons for Circle Systems of order 8

In the next Chapter we will show that every circle system with base $Z_2 \times Z_4$ is equivalent to one of the systems constructed in this Chapter. These circle systems along with the (Z_8, 0), (Z_8, 1), and (Z_8, 2) circle systems constructed in Sections 1.2, 1.5, 4.1 are pairwise nonequivalent. This can be justified by considering several parameters of a circle system which are invariant under equivalence. Besides having the same order, (number of vertices), and size, two equivalent circle systems must have the same *degree sequence.* The *degree* of a vertex X in a circle system Ω is the number of circles of Ω which contain X, and the *degree sequence* is sequence consisting of the degrees of each vertex of Ω listed in increasing order. Another invariant is the *pair degree sequence* ; if {x,y} is a pair of vertices of Ω, then the *pair degree* of {x,y}, denoted δδ{x,y}, is the number of circles of Ω which contain both x and y. The *pair degree sequence* of Ω is a sequence consisting of the pair degrees of each pair of points in Ω listed in increasing order. An interesting invariant involves the auxiliary circles of a circle system Ω. If x is a vertex of Ω, then the *tangential degree* of x is the number of auxiliary circles which are mutually tangent at x. In case Ω is a (G,g) circle system with the property that any two auxiliary circles (a,b,s,s), (c,d,s.s) with a+b+2s = c+d+2s = g, are tangent at x, then we say Ω is *tangentially complete;* otherwise Ω is *tangentially incomplete.* Two equivalent circle systems must have the same *tangential degree sequence.* The number and size of circle complexes is also an invariant under equivalence. The statistics for each of our seven circle systems of order 8 are as follows.

(i) (Z_8,0): *size*: 9
degree sequence: (4,4,4,4,5,5,5,5)
number of circle complexes: 2
tangentially complete

(i) (Z_8,1): *size:* 8
degree sequence: (4,4,4,4,4,4,4,4)
number of circle complexes: 0
tangentially complete

(ii) (Z_8,2): *size:* 10
degree sequence: (5.5,5,5,5,5,5,5)
number of circle complexes: 2
tangentially complete

(iii) $(Z_2 \times Z_4, (0,0))$
size: 10
degree sequence: (5.5,5,5,5,5,5,5)
number of circle complexes: 4
tangentially incomplete

(iv) $(Z_2 \times Z_4, (0,1))$
size: 8
degree sequence: (4,4,4,4,4,4,4,4)
number of circle complexes: 0
tangentially incomplete

(v) $(Z_2 \times Z_4, (0,2))$
size: 10
degree sequence: (5,5,5,5,5,5,5,5)
number of circle complexes: 4
tangentially complete

(vi) $(Z_2 \times Z_4, (1,0))$
size: 8
degree sequence: (4,4,4,4,4,4,4,4)
number of circle complexes: 0
tangentially incomplete

(vii) $(Z_2 \times Z_4, (1,1))$
size: 8
degree sequence: (4,4,4,4,4,4,4,4)
number of circle complexes: 0
tangentially incomplete

The above table indicates the pairwise nonequivalence of all seven pairs of order 8 circle systems with the exception of the $(Z_2 \times Z_4, (0,1))$ and $(Z_2 \times Z_4, (1,0))$ and $(Z_2 \times Z_4, (1,1))$ systems. To show the first two of these are nonequivalent we use pair degrees: the pair degree sequences of these circle systems are $(1^{10}2^{18})$, $(0^4 2^{24})$ respectively. The pair degree sequence of the $(Z_2 \times Z_4, (1,1))$ system is also $(1^{10}2^{18})$ and in fact this system is equivalent to the $(Z_2 \times Z_4, (0,1))$ system as we will show in Chapter 6. There is yet another circle system of order 8, (with base $Z_2 \times Z_2 \times Z_2$) which is not equivalent to any of the above, which we study in Chapter 7

Chapter 6

Circle Systems with Base $Z_2 \times Z_{2m}$

6.1 Equivalence among circle systems with base $Z_2 \times Z_{2m}$

Having considered in Chapter 5 the construction of circle systems with base $Z_2 \times Z_4$ we now move to consideration of the general case. To limit our focus we introduce the subject with a determination of equivalence classes. To show equivalence among circle systems with base $Z_2 \times Z_{2m}$ the following Lemmas are useful.

Lemma 41 *Let $\Gamma_{(1,0)}$ be a $(Z_2 \times Z_{2m}, (1,0))$ circle system and let $\Gamma_{(1,m)}$ be a $(Z_2 \times Z_{2m}, (1,m))$ circle system. Then $\Gamma_{(1,0)}$ is equivalent to $\Gamma_{(1,m)}$; ($\Gamma_{(1,0)} \approx \Gamma_{(1,m)}$).*

Proof Circles in $\Gamma_{(1,0)}$ have the form ((0,p),(0,q),(0,s),(1,t)) or ((1,p),(1,q),(1,s),(0,t)) where $p+q+s+t = 0$ (mod 2m). The mapping ϕ: $\Gamma_{(1,0)} \rightarrow \Gamma_{(1,m)}$ defined by $\phi(1,b) = (1,b+m)$ and $\phi(0,b) = (0,b)$ is an equivalence which maps each type of circle in $\Gamma_{(1,0)}$ to a circle in $\Gamma_{(1,m)}$, and thus $\Gamma_{(1,0)} \approx \Gamma_{(1,m)}$. □

Lemma 42 *Let $\Gamma_{(1,1)}$ be a $(Z_2 \times Z_{2m}, (1,1))$ circle system and let $\Gamma_{(0,1)}$ be a $(Z_2 \times Z_{2m}, (0,1))$ circle system, then $\Gamma_{(1,1)} \approx \Gamma_{(0,1)}$.*

Proof Let α = ((0,p),(0,q),(0,s),(1,t)) and β = ((1,p),(1,q),(1,s),(0,t)) denote the two types of circles in $\Gamma_{(1,1)}$, so $p+q+s+t = 1$ (mod 2m). The mapping ψ: $\Gamma_{(1,1)} \rightarrow \Gamma_{(0,1)}$ defined by $\psi(a,b) = (1+a, b)$ if b is odd and $\psi(a,b) = (a,b)$ if b is even is an equivalence: if {p,q,s} are all odd, then t must be even and $\psi(\alpha)$ = ((1,p),(1,q),(1,s),(1,t)) $\in \Gamma_{(0,1)}$ and $\psi(\beta)$ = ((0,p),(0,q),(0,s),(0,t)) $\in \Gamma_{(0,1)}$; If p is odd and {q,s,t} are all even, then $\psi(\alpha)$ = ((1,p),(0,q),(0,s),(1,t)) $\in \Gamma_{(0,1)}$ and $\psi(\beta)$ = ((0,p),(1,q),(1,s),(0,t)) $\in \Gamma_{(0,1)}$. If {p,q,s} are all even and t is odd, then $\psi(\alpha)$ = ((0,p),(0,q),(0,s),(0,t)) $\in \Gamma_{(0,1)}$ and $\psi(\beta)$ = ((1,p),(1,q),(1,s),(1,t)) $\in \Gamma_{(0,1)}$. All the possibilities have been considered so we can conclude that $\Gamma_{(1,1)} \approx \Gamma_{(0,1)}$. □

With the completion of Lemmas 41 and 42 we are prepared to give a precise description of the equivalence classes of circle systems with base $Z_2 \times Z_{2m}$.

Theorem 43 *Let Ω be a circle system with base $Z_2 \times Z_{2m}$. (i) if m is odd then $\Omega \approx \Gamma_{(0,0)}$ or $\Omega \approx \Gamma_{(0,1)}$ and $\Gamma_{(0,0)} \not\approx \Gamma_{(0,1)}$; (ii) If $m = 2q$ is even, then Ω has base $Z_2 \times Z_{4q}$. If q is odd then Ω is equivalent to one of $\{\Gamma_{(0,0)}, \Gamma_{(0,1)}, \Gamma_{(0,2)}, \Gamma_{(1,0)}\}$ and these circle systems are pairwise nonequivalent; (iii) If q is even, say $q = 2t$, then Ω has base $Z_2 \times Z_{8t}$ and Ω is equivalent to one of $\{\Gamma_{(0,0)}, \Gamma_{(0,1)}, \Gamma_{(0,2)}, \Gamma_{(1,0)}, \Gamma_{(1,2)}\}$ and these circle systems are pairwise nonequivalent. Moreover, no circle system with base Z_{4m} is equivalent to any circle system with base $Z_2 \times Z_{2m}$.*

Proof Suppose m is odd and $t \in Z_{2m}$. Let $\Gamma_{(0,t)}$ and $\Gamma_{(1,t)}$ denote $(Z_2 \times Z_{2m}, (0,t))$ and $(Z_2 \times Z_{2m}, (1,t))$ circle systems respectively. Let $k \in Z_{2m}$, then $\phi: Z_2 \times Z_{2m} \rightarrow Z_2 \times Z_{2m}$ defined by $\phi(x) = x + (0,k)$ is an equivalence between $\Gamma_{(0,0)}$ and $\Gamma_{(0,4k)}$. Since m is odd the subgroup $\langle 4 \rangle$ of Z_{2m} generated by 4 contains all the even values in Z_{2m}, so $\Gamma_{(0,0)} \approx \Gamma_{(0,t)}$ for every even value of t in Z_{2m}. The same mapping ϕ is also an equivalence between $\Gamma_{(0,1)}$ and $\Gamma_{(0,4k+1)}$ with the consequence that $\Gamma_{(0,1)} \approx \Gamma_{(0,t)}$ for every odd value of t in Z_{2m}. Again, ϕ is an equivalence between $\Gamma_{(1,0)}$ and $\Gamma_{(1,4k)}$, and also between $\Gamma_{(1,1)}$ and $\Gamma_{(1,4k+1)}$ with the consequence that $\Gamma_{(1,0)} \approx \Gamma_{(1,t)}$ for every even t, and $\Gamma_{(1,1)} \approx \Gamma_{(1,t)}$ for every odd t in Z_{2m}. By Lemma 41, $\Gamma_{(1,0)} \approx \Gamma_{(1,m)}$, and since m is odd, we have $\Gamma_{(1,m)} \approx \Gamma_{(1,1)}$. By Lemma 42 we have $\Gamma_{(1,1)} \approx \Gamma_{(0,1)}$. It follows that $\Omega \approx \Gamma_{(0,0)}$ or $\Omega \approx \Gamma_{(0,1)}$.

Now suppose m = 2q is even and q is odd, then the mapping ϕ produces the following equivalences : $\Gamma_{(0,0)} \approx \Gamma_{(0,4t)}$; $\Gamma_{(0,1)} \approx \Gamma_{(0,4t+1)}$; $\Gamma_{(0,2)} \approx \Gamma_{(0,4t+2)}$; $\Gamma_{(0,3)} \approx \Gamma_{(0,4t+3)}$; $\Gamma_{(1,0)} \approx \Gamma_{(1,4t)}$; $\Gamma_{(1,1)} \approx \Gamma_{(1,4t+1)}$; $\Gamma_{(1,2)} \approx \Gamma_{(1,4t+2)}$; $\Gamma_{(1,3)} \approx \Gamma_{(1,4t+3)}$ for $t \in Z_{2m}$. Thus Ω is equivalent to one of $\Gamma_{(0,0)}, \Gamma_{(0,1)}, \Gamma_{(0,2)}, \Gamma_{(0,3)}, \Gamma_{(1,0)}, \Gamma_{(1,1)}, \Gamma_{(1,2)}, \Gamma_{(1,3)}$. By Lemma 41, $\Gamma_{(1,0)} \approx \Gamma_{(1,m)}$, and since m is twice an odd number, $\Gamma_{(1,m)} \approx \Gamma_{(1,2)}$, thus $\Gamma_{(1,0)} \approx \Gamma_{(1,2)}$. Now the mapping $\psi: Z_2 \times Z_{2m} \rightarrow Z_2 \times Z_{2m}$ defined by $\psi(a,b) = (a, -b)$ is an equivalence and thus $\Gamma_{(0,1)} \approx \Gamma_{(0,-1)} \approx \Gamma_{(0,4t+3)} \approx \Gamma_{(0,3)}$ for t = -1, and similarly $\Gamma_{(0,1)} \approx \Gamma_{(1,1)} \approx \Gamma_{(1,3)}$. thus (ii) is proved, (except for the nonequivalence). Now suppose q is even. then the mapping ϕ shows again that Ω must be equivalent to one of circle systems : $\Gamma_{(0,0)}, \Gamma_{(0,1)}, \Gamma_{(0,2)}, \Gamma_{(0,3)}, \Gamma_{(1,0)}, \Gamma_{(1,1)}, \Gamma_{(1,2)}, \Gamma_{(1,3)}$. Also as above, $\Gamma_{(0,1)} \approx \Gamma_{(0,3)}$ and $\Gamma_{(0,1)} \approx \Gamma_{(1,1)} \approx \Gamma_{(1,3)}$, so we have (iii), except for the nonequivalence.

To prove nonequivalence, first note that $\Gamma_{(0,0)}, \Gamma_{(0,2)}$ are the only circle systems which contain circle complexes, so they cannot be equivalent to any of the other circle systems with base $Z_2 \times Z_{2m}$. In Chapter 10 it is shown, with n = 2m, that the size of $\Gamma_{(0,0)}$ is $(1/3)(n-1)(n^2 - 2n + 3)$ if m is odd, and the size of $\Gamma_{(0,0)}$ is $(1/3)(n-2)(n^2 - n + 3)$ if m is even. Also shown in Chapter 10 is that the size of $\Gamma_{(0,2)}$ is $(1/3)(n)(n^2 - 3n + 5)$ if m is even. These size differences allow us to conclude that for m even, $\Gamma_{(0,0)} \not\approx \Gamma_{(0,2)}$. The size of a $(Z_2 \times Z_{2m}, (1,y))$ circle system is given by $2\binom{n}{3} = (1/3)(n)(n-1)(n-2)$. This formula is easy to understand: the correspondence:

((0,a),(0,b),(0,c),(1,d))→((1,a),(1.b),(1,c), (0,d))

between circles in $\Gamma_{(1,y)}$ is one – one and there are $\binom{n}{3}$ circles of the type ((0,a),(0,b),(0,c),(1,d)). In Chapter 10 it is also determined that the size of $\Gamma_{(0,1)}$ is given by $(1/3)(n)(n-1)(n-2)$, thus size cannot be used to distinguish between $\Gamma_{(0,1)}, \Gamma_{(1,0)}, \Gamma_{(1,1)}, \Gamma_{(1,2)}$. Suppose $8|n$ and $\varphi: \Gamma_{(1,0)} \rightarrow \Gamma_{(1,2)}$ is an equivalence. Consider:

(i) $(1,1) + (0,-1) + (0,0) + (0,0) = (1,0)$

(ii) $(1,1) + (0,-1) + (1,0) + (1,0) = (1,0)$

(iii) $(0,0) + (0,0) + (0,0) + (1,0) = (1,0)$

(iv) $(0,0) + (1,0) + (1,0) + (1,0) = (1,0)$

So if φ (1,1) = (a,b); $\phi(0,-1) = (c,d)$; $\phi(0,0) = (e,f)$; $\phi(1,0) = (g,h)$; Then we must have:

(v) $(a,b) + (c,d) + (e,f) + (e,f) = (a,b) + (c,d) + (0,2f) = (1,2)$

(vi) $(a,b) + (c,d) + (g,h) + (g,h) = (a,b) + (c,d) + (0,2h) = (1,2)$

(vii) $(e,f) + (e,f) + (e,f) + (g,h) = (e,3f) + (g,h) = (1,2)$

From these equations we obtain:

(viii) $2f = 2h,$

(ix) $3f + h = 2.$

If $f = h$, then (ix) yields $4f = 2 \bmod n$, which is not possible since $4|n$. So the only solution for (viii) has the form $h = f + 4q$. Then (ix) yields $4f + 4q = 2 \bmod n$ which, again, has no solutions since $4|n$. We can show similarly that $\Gamma_{(1,0)}$ and $\Gamma_{(0,1)}$ are not equivalent if n is a multiple of 4. From (iii), (iv) above, and assuming an equivalence we obtain:

(x) $(e,3f) + (g,h) = (0,1)$

(xi) $(e,f) + (g,3h) = (0,1)$

From these we obtain $3f+h = 1$ and $f+3h = 1$, and adding these last two equations yields $4(f+h) = 2 \bmod n$. But since n is a multiple of 4 this is impossible. The same argument shows that $\Gamma_{(1,0)} \not\approx \Gamma_{(1,1)}$. To show $\Gamma_{(1,2)} \not\approx \Gamma_{(0,1)}$ if $n = 4q$ is a multiple of 4, suppose $f\colon \Gamma_{(1,2)} \rightarrow \Gamma_{(0,1)}$ is an equivalence and suppose $f(1,1) = (a,b)$; $f(0,1) = (c,d)$; $f(0,0) = (e,f)$ and $f(1,0) = (g,h)$. Since $(0,0) + (0,0) + (0,1) + (1,1) = (1,2)$, we must have:

(xii) $(e,f) + (e,f) + (c,d) + (a,b) = (0,1)$ *which implies* $2f + b + d = 1.$

From $(0,0) + (1,0) + (0,1) + (0,1) = (1,2)$ we obtain:

(xiii) $(e,f) + (g,h) + (c,d) + (c,d) = (0,1)$ *which implies* $f+h+2d = 1.$

And from $(0,0) + (1,0) + (1,1) + (1,1) = (1,2)$ we obtain:

(xiv) $(e,f) + (g,h) + (a,b) + (a,b) = (0,1)$ *which implies* $f+h+2b = 1.$

From (xiv), (xiii) we obtain $2b = 2d$. If $b = d$ then (xii) yields $2f + 2b = 1 \bmod n$ which is impossible since $4|n$. Then we must have $d = b + 2q$, but then (xii) yields $2f + 2b + 2q = 1$ which is also impossible. We conclude that $\Gamma_{(1,2)} \not\approx \Gamma_{(0,1)}$ if $4|n$.

It remains only to prove the last statement of the Theorem. In the table below we compare sizes of circle systems Ω with base $Z_2 \times Z_{2m}$ or Z_{4m}.

Ω	size
$(Z_{4m}, 0)$	$(1/3)(m-1)(8m^2 - 4m + 3)$
$(Z_{4m}, 2)$	$(1/3)(m)(8m^2 - 12m + 7)$
$(Z_{4m}, 1)$	$(4/3)(m)(m-1)(2m-1)$

$(Z_2 \times Z_{2m}, (1,0))$	$(4/3)(m)(m-1)(2m-1)$
$(Z_2 \times Z_{2m}, (1,2))$	$(4/3)(m)(m-1)(2m-1)$
$(Z_2 \times Z_{2m}, (0,0))$, m odd	$(1/3)(2m-1)(4m^2 - 4m + 3)$
$(Z_2 \times Z_{2m}, (0,0))$, m even	$(2/3)(m-1)(4m^2 - 2m + 3)$
$(Z_2 \times Z_{2m}, (0,1))$	$(4/3)(m)(m-1)(2m-1)$
$(Z_2 \times Z_{2m}, (0,2))$	$(2/3)(m)(4m^2 - 6m + 5)$

The justification for these formulas can be found in Chapter 10. The Table only fails to distinguish between the $(Z_{4m}, 1)$ and the $(Z_2 \times Z_{2m}, (1,0))$, $(Z_2 \times Z_{2m}, (1,2))$, and $(Z_2 \times Z_{2m}, (0,1))$ circle systems. To show nonequivalence we consider pair degree sequences in the following Table. Recall that if {x,y} is a pair of vertices of a circle system Ω, then the *pair degree sequence* denoted δδ{x,y} is a sequence consisting of the *pair degrees* of each pair of points in Ω, where the pair degree of {x,y} is the number of circles in Ω which contain both x and y.

Ω	pair degree sequence
$(Z_{4m}, 1)$	$(2m-2)^{4m^2+2m}(2m-3)^{4m^2-4m}$
$(Z_2 \times Z_{2m}, (0,1))$	$(2m-2)^{6m^2-2m}(2m-3)^{4m}(2m-4)^{2m^2-4m}$
$(Z_2 \times Z_{2m}, (1,2))$	$(2m-2)^{7m^2-2m}(2m-3)^{4m}(2m-4)^{m^2-4m}$
$(Z_2 \times Z_{2m}, (1,0))$,m even	$(2m-2)^{6m^2-2m+8}(2m-3)^{4m-16}(2m-4)^{2m^2-4m+8}$
$(Z_2 \times Z_{2m}, (1,0))$,m odd	$(2m-2)^{6m^2-2m+4}(2m-3)^{4m-8}(2m-4)^{2m^2-4m+4}$

The determination of these formulas can also be found in Chapter 10. Since the $(Z_{4m}, 1)$ circle system contains no vertices with degree $2m - 4$, and since $2m^2 - 4m + 8 = 0$ and $2m^2 - 4m + 4 = 0$ have no real solutions, we can conclude that the $(Z_{4m}, 1)$ circle system is not equivalent to a $(Z_2 \times Z_{2m}, (1,0))$ circle system. Setting $2m^2 - 4m = 0$ yields $m = 2$, so there is only the possibility that the $(Z_8, 1)$ circle system is equivalent to the $(Z_2 \times Z_4, (0,1))$ system, but these we have already distinguished by means of tangential completeness. Setting $m^2 - 4m = 0$ yields $m = 4$, so there is only the possibility that circle systems $(Z_{16}, 1)$ and $(Z_2 \times Z_8, (1,2))$ are equivalent. But when 4 is substituted for the pair degree sequence formulas for these two circle systems, we obtain very different expressions. □

6.2 The $(Z_2 \times Z_{2m}, (0,0))$ circle system

Having considered questions of equivalence, we will now determine the existence of general circle systems with base $Z_2 \times Z_{2m}$. Similar to Theorem 19 and the opening paragraph of Section 4.1, we can show that a $(Z_2 \times Z_{8m}, (0,0))$ circle system contains a $(Z_2 \times Z_{2m}, (0,1))$ and a $(Z_2 \times Z_{2m}, (0,2))$ circle system as subsystems.

Theorem 44 *Let Ω be a $(Z_2 \times Z_{4m}, (0,0))$ circle system. Then a $(Z_2 \times Z_{2m}, (0,2))$ circle system occurs as a subsystem of Ω. Let Γ be a $(Z_2 \times Z_{8m}, (0,0))$ circle system. Then a $(Z_2 \times Z_{2m}, (0,1))$ circle system occurs as a subsystem of Γ. Let Λ be a $(Z_2 \times Z_{4m}, (1,0))$ circle system. Then a $(Z_2 \times Z_{2m}, (1,2))$ circle system occurs as a subsystem of Λ.*

Proof Let W denote the set of points of Ω of the form (s, 2t – 1). These points form a closed subset of the circles of Ω. If the points in W are relabeled according to the rule (s, 2t – 1)→(s,t) we obtain a $(Z_2 \times Z_{2m}, (0,2))$ circle system. Let U denote the set of points of Γ of the form (s, 4t-1). These points form a closed subset of the circles of Γ. If the points of U are relabeled according to the rule (s, 4t-1)→(s,t), a $(Z_2 \times Z_{2m}, (0,1))$ circle system is obtained. Finally, let V denote the set of points of Λ of the form (s, 2t – 1), and relabel these by the rule (s, 2t – 1)→(s,t) to obtain a $(Z_2 \times Z_{2m}, (1,2))$ circle system. □

From Theorem 44 we see that the existence and construction of any circle system with base $Z_2 \times Z_{2m}$ will be proved once the existence and construction of the $(Z_2 \times Z_{2m}, (0,0))$ and the $(Z_2 \times Z_{2m}, (1,0))$ circle systems is proved. The following general result will be useful.

Theorem 45 *Let Ω be a (G,0) and Γ and a (H,0) circle system each with the same irreducible self-inversive cubic envelope γ. Suppose also that Ω, Γ have only the point with label 0 in common. Relabel each point g of Ω with the ordered pair (g,0) and relabel each point h of Γ with the ordered pair (0,h). Let (γ, δ) be the THA defined on γ, and for $g \in G$ and $h \in H$ let $(g,h) = \delta((0,0),(-g,0),(0,-h))$. Then the set of points so defined yields a $(G \times H, (0,0))$ circle system on γ.*

Proof Let $(g_1, h_1), (g_2, h_2), (g_3, h_3), (g_4, h_4) \in G \times H$ with $(g_1, h_1)+(g_2, h_2)+(g_3, h_3)+(g_4, h_4) = (0,0)$. Recall from Theorem 16 that the ternary operation δ is defined on γ by setting $\delta(a,b,c)$ equal to the fourth point besides a, b, c which lies on γ and on circle (a,b,c). Thus $a + b + c + \delta(a,b,c) = (0,0)$. In the following derivation the ternary hypercommutative property of δ is employed:

$$\delta((g_1, h_1), (g_2, h_2), (g_3, h_3)) =$$

$$\delta\{\delta((0,0),(-g_1,0),(0,-h_1)), \delta((0,0),(-g_2,0),(0,-h_2)), \delta((0,0),(-g_3,0),(0,-h_3))\} =$$

$$\delta\{\delta((0,0),(0,0),(0,0)), \delta((-g_1,0),(-g_2,0),(-g_3,0)), \delta((0,-h_1),(0,-h_2),(0,-h_3))\} =$$

$$\delta\{(0,0), (g_1 + g_2 + g_3, 0), (0, h_1 + h_2 + h_3)\} = (-(g_1 + g_2 + g_3), -(h_1 + h_2 + h_3)) = (g_4, h_4). \ \square$$

In light of Theorem 45 we can construct a $(Z_2 \times Z_{2m}, (0,0))$ circle system on a type IV self-inversive cubic curve γ by first constructing a $(Z_{2m}, 0)$ circle system on the bell. By Corollary 5 there exists a circle w_0 which inverts all pairs of points of the form {t,-t} and which contains the points 0, m. This circle, in fact, inverts the entire curve γ onto itself and has center X on the bell. If $g \in Z_{2m}$, then we relabel by g → (0,g), thus the points (0,0) and (0,m) lie on the bell and on circle w_0. A $(Z_2, 0)$ circle system has only the two points 0,1 which we relabel by 0 →(0,0), 1 →(1,0). We put the point (1,0) on the oval, so that the circle systems $(Z_{2m}, 0)$ and $(Z_2,0)$ have only the point 0 = (0,0) in common, as per the requirement of Theorem 44. The requirement $\delta((1,0),(1,0),(1,0)) = (1,0)$ forces the point (1,0) on the oval to be one of the two points on the oval which square to X. We relabel $w_0 = w_{(0,0)}$ since this circle will invert all pairs in the resulting $(Z_2 \times Z_{2m}, (0,0))$ circle system which sum to (0,0). Thus the points (1,0), (1,m) lie on the oval and on circle $w_{(0,0)}$. The $(Z_{2m}, 0)$ circle system can be constructed on the bell using the technique described in Theorem 9, or we can construct a perfect 2m-gon on the bell with $X_0 = X$ and apply Theorem 34. With the relabeling described in Theorem 45 we actually obtain a perfect $Z_2 \times Z_{2m}$ – gon on γ with $X_{(0,0)} = X$ which is also our desired $(Z_2 \times Z_{2m}, (0,0))$ circle system. In Figure 41 we illustrate a $(Z_2 \times Z_{8,}\ (0,0))$ circle system constructed in this manner.

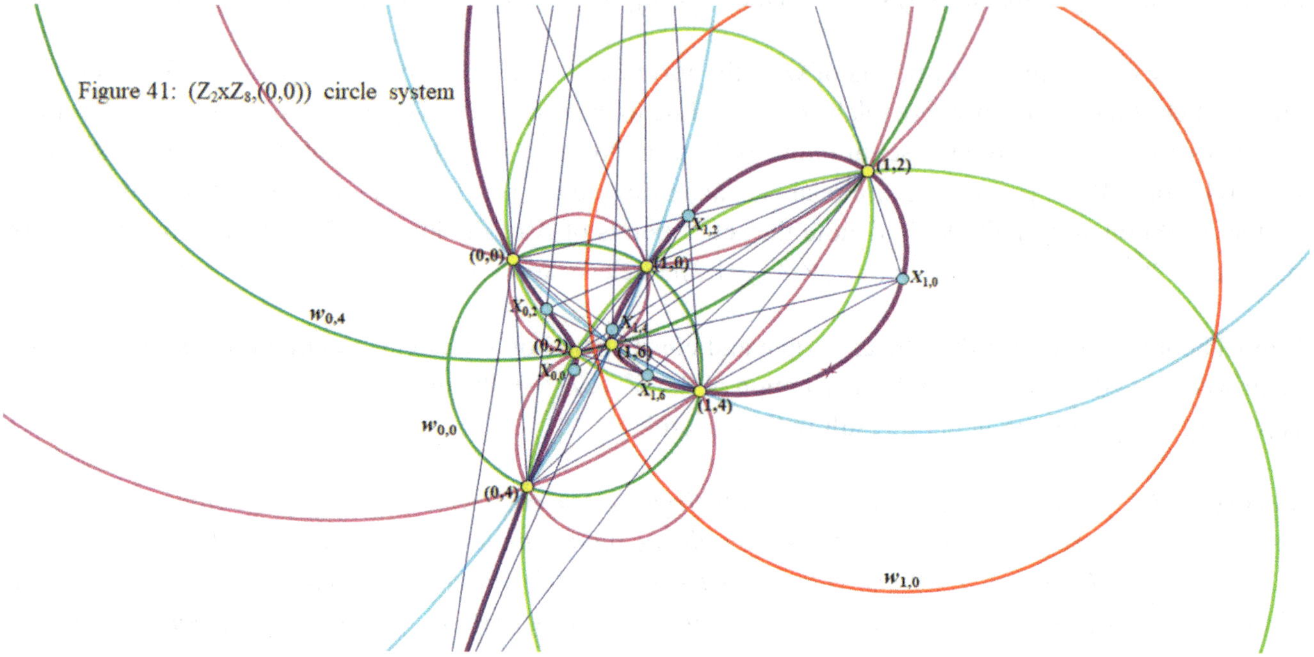

Figure 41: $(Z_2xZ_8,(0,0))$ circle system

6.3 The $(Z_2 \times Z_{2m}, (1,0))$ circle system

Let $\Gamma_{(1,0)}$ denote a $(Z_2 \times Z_{2m}, (1,0))$ circle system. By Lemma 41, $\Gamma_{(1,0)} \approx \Gamma_{(1,m)}$ and if m is odd then $\Gamma_{(1,m)} \approx \Gamma_{(1,1)}$. Then by Lemma 42, $\Gamma_{(1,1)} \approx \Gamma_{(0,1)}$. So, if m is odd then $\Gamma_{(1,0)} \approx \Gamma_{(0,1)}$, and by Theorem 44 we can obtain $\Gamma_{(1,0)}$ as a subsystem of a $(Z_2 \times Z_{8m}, (0,0))$ circle system. If m is even then the only $(Z_2 \times Z_{2m}, (1,0))$ circle system we have so far constructed is with m = 2 and is illustrated in Figure 38. <u>We will show that a $(Z_2 \times Z_{2m}, (1,0))$ circle system with m even and $m \geq 4$ *cannot* have a self-inversive cubic curve as an envelope.</u> Since all circle systems so far constructed, with the exception of Figure 38, have envelopes which are either self-inversive cubic curves or the inverse of such a curve, we doubt the existence of $(Z_2 \times Z_{2m}, (1,0))$ circle systems with m even and $m \geq 4$. To establish the underlined statement a close look at the algebraic structure of a self-inversive cubic curve will be undertaken.

Suppose a $(Z_2 \times Z_8, (1,0))$ circle system Ω lies on a self-inversive cubic curve γ, and let (γ,δ) be the THA defined on γ. The vertices of Ω can be partitioned into eight two-element subalgebras of (γ,δ) namely {(0,0),(1,0)}, {(0,1),(1,5)}, {(1,1),(0,5)}, {(0,2),(1,2)}, {(0,3),(1,7)}, {(1,3),(0,7)}, {(0,4),(1,4)}, {(0,6), (1,6)}. We will determine exactly where these two-element subalgebras lie on a self-inversive cubic curve.

Theorem 46 *Let γ be a self-inversive Type IV cubic curve and let X,Y be points on the bell which serve as centers of circles w, v respectively, which invert γ onto itself. Let (γ,δ) denote the THA on γ, and suppose {a,b} is a two element subalgebra of (γ,δ) with $\delta(a,a,a) = b$ and $\delta(b,b,b) = a$ and $a \neq b$. Then: (i) $(a*a)*(a*a) = X*Y = (b*b)*(b*b)$; (ii) $a*a = b*b$; (iii) there exist exactly four such pairs {a,b} on γ, two of which occur on the oval and two on the bell; (iv) If {a,b},{c,d} denote the two pairs which lie on the oval (bell), then they form a subalgebra of the oval (bell) in which $\delta(a,b,c) = d$, $b*d = a*c$, $b*c = a*d$ and $\{b*d, a*d\} = \{X,Y\}$.*

Proof Let {a,b} be a two element subalgebra of (γ,δ) with δ(a,a,a) = b, δ(b,b,b) = a, and a≠b. Then by Theorem 16, δ(a,a,a) = X* {(X*a)*(a*a)}= b and thus X*b = (X*a)*(a*a). Similarly, X*a = (X*b)*(b*b). Then X*b = {(X*b)*(b*b)}*(a*a) from which it follows from the BH axiom p*(p*q) = q, and by cancellation that a*a = b*b. A basic fact that will be used in the following derivation is that X*X = ∞ where ∞ denotes the unique point at infinity on a self-inversive cubic curve γ and X can be the center of any of the circles which invert or antiinvert γ onto itself. Consider:

$$b*b = \delta(a,a,a)*\ \delta(a,a,a) = \{X*\{(X*a)*(a*a)\}*\ \{X*\{(X*a)*(a*a)\} =$$

$$\infty\ *\ \{(\infty*(a*a))*((a*a)*(a*a)\},$$

$$\text{so } a = \delta(b,b,b) = X*\{(X*b)*(b*b)\} = X*\ \{\{(X*a)*(a*a)\}*\{\infty\ *\ \{(\infty*(a*a))*((a*a)*(a*a)\}\},$$

$$a = (X*\ \infty)\ *\{\{(X*a)*(a*a)\}*\{\infty\ *\ \{(\infty*(a*a))*((a*a)*(a*a)\}\},$$

$$a = \{X*((X*a)*(a*a))\}*\ \{(\infty*(a*a))*((a*a)*(a*a)\},$$

$$a = \{(X*\ \infty*((X*a)*(a*a))\}*\ \{(\infty*(a*a))*((a*a)*(a*a)\},$$

$$a = \{(a*((\infty*(a*a))\}*\ \{(\infty*(a*a))*((a*a)*(a*a)\},$$

$$\infty*(a*a) = \{(\infty*(a*a))*((a*a)*(a*a)\},$$

$$\{(a*a)*(a*a)\}*\{(a*a)*(a*a)\} = \infty*\ \infty.$$

The point ∞*∞ has four square-roots on γ, namely {∞, X*Y, X*Z, R*Y} where Z, R lie on the oval and are centers of circles which invert and antiinvert respectively, γ onto itself. If Q = (a*a)*(a*a) then from above we have Q*Q = ∞*∞, and thus Q ∈ {∞, X*Y, X*Z, R*Y}. The points X*Z, R*Y lie on the oval and thus cannot be squares. Since Q is a square we must have Q∈{∞, X*Y}. If Q = ∞, then a*a ∈ {X,Y,Z,R} since these are precisely the four square-roots of ∞. Since R, Z lie on the oval, they cannot be squares, so we must have a*a ∈ {X,Y}. If a*a = X, then δ(a,a,a) = X*{(X*a)*(a*a)} = X*a = a which we do not allow. Similarly if a*a = Y, then δ(a,a,a) = Y*{(Y*a)*(a*a)} = Y*a = a. Consequently we must have Q = X*Y, i.e., (a*a)*(a*a) = X*Y. The point X*Y has four square-roots, two on the bell and two on the oval, but only the two on the bell are squares, so a*a must be one of these two, and these two each have, in turn, four square-roots, two on the bell and two on the oval, as in Figure 42. We have proved (i), (ii) and (iii).

To prove (iv) suppose {a,b}, {c,d} are two pairs on the oval with δ(a,a,a) = b, δ(b,b,b) = a, and a≠b, and δ(c,c,c) = d, δ(d,d,d) = c, and c≠d. Let U,T denote the two square-roots of X*Y which lie on the bell and suppose a*a = U and d*d = T. Since a*a = b*b we have (a*a)*(b*b) = X*Y and thus (a*b)*(a*b) = X*Y. Since a,b lie on the oval, a*b lies on the bell and thus a*b is one of the two square-roots of X*Y on the bell. So a*b ∈ {U,T}. If a*b = U = a*a, we obtain a =b, so we must have a*b = T = d*d. Similarly c*d = U = a*a. Also c*c = d*d follows from (ii) with c,d in place of a,b. Consider: δ(a,b,c) = X*{(X*c)*(a*b)} = X*{(X*c)*(d*d)} = δ(c,d,d) = d.

To establish (iv) it remains to show that such expressions as δ(a,c,c), δ(b,d,d), etc., lie in {a,b,c,d}. Consider: δ(a,c,c) = X*((X*a)*(c*c)) = X* ((X*a)*(a*b)) = δ(a,a,b) = a and δ(b,d,d)= X*((X*b)*(d*d)) = X*((X*b)*(a*b)) = δ(a,b,b) = b. Showing that δ(a,d,d) = a and δ(b,c,c) = b is accomplished similarly. Thus {a,b,c,d} is a subalgebra of (γ,δ). The last part of (iv) is proved as follows: consider: $(a*c)*(b*c) = (a*b)*(c*c) = T*T = X*Y$, so $a*c = (b*c)*(X*Y) = (b*c)*(U*U) = (b*U)*(c*U) = b*d$, and similarly $a*d = b*c$. Finally, $(b*d)*(b*d) = (b*b)*(d*d) = U*T$. Now $(U*T)*(U*T) = (U*U)*(T*T) = (X*Y)*(X*Y) = (X*X)*(Y*Y) = \infty*\infty$ and thus $U*T$ is a square root of $\infty*\infty$ which lies on the bell. As in the first part of the proof, we must have $U*T \in \{\infty, X*Y\}$. If $U*T = X*Y$ then $U*T = T*T$ yields $U = T$ by cancellation, but this is not the case. So we must have $U*T = \infty$. Then we have $(b*d)*(b*d) = \infty$, and since $b*d$ lies on the bell, it must be that $b*d \in \{X,Y\}$. Similarly $a*d \in \{X,Y\}$ and since a, b are not equal we must have that one of $\{a*d, b*d\}$ is X and the other equals Y. The same result is obtained if we suppose that $\{a,b\}$ and $\{c,d\}$ lie on the bell. □

We have seen that a $(Z_2 \times Z_{8,}$ (1,0)) circle system which lies on a self-inversive cubic curve γ has a total of eight pairs of points $\{a,b\}$ which form the type of two-element subalgebra described in Theorem 46. But by Theorem 46 there exist only four such two element subalgebras on γ, so we can conclude that a $(Z_2 \times Z_{8,}$ (1,0)) circle system cannot lie on a self-inversive cubic. We will argue that no $(Z_2 \times Z_{4q,}$ (1,0)) circle system Ω can have a self-inversive cubic envelope for any positive integer q.

Figure 42: two-element subalgebras

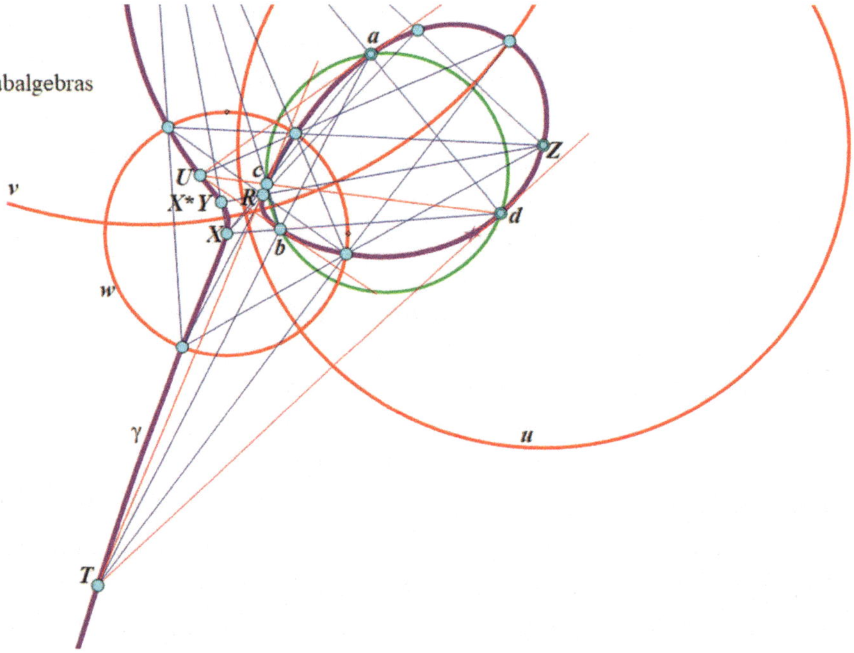

Theorem 47 *A $(Z_2 \times Z_{4q,}$ (1,0)) circle system Ω cannot have a self-inversive cubic envelope for any positive integer q.*

Proof Suppose Ω has a self-inversive cubic envelope γ, and let (γ,δ) denote the THA defined on γ. The points of any circle system including Ω form a subalgebra of (γ,δ); let (Ω,δ) denote the subalgebra formed by the points of our $(Z_2 \times Z_{4q,}$ (1,0)) circle system. It is convenient to distinguish between the two types of two-element subalgebras of (γ,δ) : Type I consists of two points $\{a,b\}$ with $\delta(a,a,a) = a$ and $\delta(b,b,b) = b$ and $\delta(a,a,b) = b$, and Type II, (the one under discussion in Theorem 46), has $\delta(a,a,a) = b$ and $\delta(b,b,b) = a$. Our circle system Ω has the following four Type II subalgebras: $\{(0,0),(1,0)\}$, $\{(0,q),(1,q)\}$, $\{(0,2q), (1,2q)\}$, $\{(0,3q),(1,3q)\}$. A combination of any two of these pairs does not produce a circle in Ω, contrary to part (iv) of Theorem 46, so we can conclude that Ω cannot have a self-inversive cubic curve as envelope. □

Note that by Theorem 47, even the $(Z_2 \times Z_4, (1,0))$, which exists and is illustrated in Figure 38, cannot have a self-inversive cubic envelope. On the basis of the result of Theorem 47, and the rarity of circle systems without a self-inversive, (or inverse of a self-inversive), cubic envelope, we make the following Conjecture.

Conjecture 48 *If q is an integer greater than 1, then a $(Z_2 \times Z_{4q},(1,0)$ circle system does not exist.*

6.4 The $(Z_2 \times Z_{8q,}$ (1,2)) circle system

In this Section we take a similar approach to that taken in Section 6.3, i.e., we identify a subalgebra structure which must exist in a $(Z_2 \times Z_{8q}, (1,2))$ circle system, and then show that a self-inversive cubic curve cannot support such a structure. We begin with the case q = 1.

Let Ω be a $((Z_2 \times Z_{8,}$ (1,2)) circle system, and suppose the vertices of Ω lie on a self-inversive cubic curve γ. We find that the 16 vertices of Ω form the following four four point subalgebras of the THA (γ,δ).

{(0,0),(1,2),(0,4),(1,6)}, {(0,2),(1,4),(0,6),(1,0},

{(0,1),(1,7),(0,5),(1,3)}, {(0,3),(1,1),(0,7),(1,5)}.

If we define φ: (γ,δ)⟶ (γ,δ) by φ(a) = δ(a,a,a) and t is the least positive integer such that $\varphi^t(a) = a$, then we call the set of points $\{a, \varphi(a), \varphi^2(a), \ldots, \varphi^{t-1}(a)\}$ the *t- cycle generated by a.* Each of the above subalgebras is a 4-cycle, so it becomes important to determine the number and character of the four cycles in (γ,δ).

We will start by assuming that γ is a Type IV self-inversive cubic curve. Let w, v, u be the inverting circles with centers X,Y,Z respectively, as in Figure 42, and let a be a point on the oval. If φ(a) = a, then X*{(X*a)*(a*a)} = a implies (a*a)*(a*a) = ∞ which implies, in turn, that a*a ∈ {X,Y}. there are two points on the oval which square to X and two points on the oval which square to Y. These are precisely the four points p,q,r,s of Figure 43, where the circles w, v meet the oval. If $\varphi^2(a) = a$, then we obtain, as in the proof of Theorem 46: {(a*a)*(a*a)}*{(a*a)*(a*a)} = ∞*∞ which gave us a*a ∈ {X,Y,U,T} where U,T are the two square-roots of X*Y on the bell. The eight choices for a on the oval are the points labeled {p,q,r,s, a,b,c,d} of Figure 43. Here, {a,b} and {c,d} constitute the only 2-cycles on the oval, and these are also subalgebras of (γ,δ).

As a point *a* makes one circuit of the oval, the point φ(a) = δ(a,a,a) makes three circuits in the reverse direction. If *a* lies on the oval, let μ_t denote the number of distinct solutions to $\varphi^t(a) = a$, and let a *t - section* denote the section of the oval between two consecutive solutions. As point *a* moves across a t-section, the point φ(a) moves across μ_1 -1 t-sections in the opposite direction. Thus the total number of 1-sections crossed by φ(a) as *a* makes one circuit of the oval is given by $(\mu_1 -1)\ \mu_1$, and this must equal $3\mu_1$ which yields $\mu_1 = 4$, in agreement with our observations above. Now $\varphi^2(a) = \delta(\varphi(a), \varphi(a), \varphi(a))$ makes 3 circuits of the oval in the opposite direction as φ(a) makes one circuit, and consequently $\varphi^2(a)$ makes nine complete circuits of the oval, in the *same* direction, as *a* makes one circuit. Since a, $\varphi^2(a)$ move in the same direction, $\varphi^2(a)$ will cover $\mu_2 + 1$ 2-sections as *a* covers one 2-section. Thus, as *a* makes one circuit around the oval, $\varphi^2(a)$ will cross $(\mu_2 + 1)\ \mu_2$ 2-sections. Then we must have $(\mu_2 + 1)\ \mu_2 = 9\mu_{2,}$ which yields $\mu_2 = 8$, again in agreement with our previous observations.

We proceed in the same way to compute the number of solutions to $\varphi^3(a) = a$, for a on the oval. In this case $\varphi^3(a)$ makes 27 circuits of the oval in the opposite direction, as *a* makes one circuit. As *a* crosses one 3-section, $\varphi^3(a)$ crosses $(\mu_3 - 1)$ 3-sections in the opposite direction, so as *a* makes a full circuit of the oval, $\varphi^3(a)$ crosses $(\mu_3 - 1)\,\mu_3$ 3-sections. Then we have $(\mu_3 - 1)\,\mu_3 = 27\mu_3$ which yields $\mu_3 = 28$. This count includes the four solutions to $\varphi(a) = a$, and the remaining 24 solutions form eight 3-cycles. Suppose $\{a, b = \varphi(a), c = \varphi^2(a)\}$ is one such 3-cycle. Then $\delta(a,b,c) = \delta(\varphi(c), \varphi(a), \varphi(b)) = \delta(\delta(c,c,c), \delta(a,a,a), \delta(b,b,b)) = \delta(\delta(a,b,c), \delta(a,b,c), \delta(a,b,c))$, which shows that $\delta(a,b,c)$ is δ- idempotent and thus belongs to no 3-cycle. So, in fact, no 3-cycle is a subalgebra of (γ,δ).

Finally, suppose *a* belongs to the oval and $\varphi^4(a) = a$. Now as *a* makes one circuit of the oval, $\varphi^4(a)$ makes 81 circuits in the same direction, and as *a* crosses one 4-section, $\varphi^4(a)$ crosses $\mu_4 + 1$, 4-sections. So, as *a* makes a full circuit of the oval, $\varphi^4(a)$ crosses $(\mu_4 + 1)\mu_4$ 4-sections. We then have $(\mu_4 + 1)\mu_4 = 81\mu_4$ yielding $\mu_4 = 80$. If we now subtract the 8 solutions to $\varphi^2(a) = a$, (which are also solutions to $\varphi^4(a) = a$), we obtain a total of 72 solutions which form 18 4-cycles. There are thus 18 4-cycles on the oval of a Type IV self-inversive cubic. If we construct a $(Z_{80}, 0)$ circle system on the oval, we find that it is comprised precisely of all 18 4-cycles as well as the 8 solutions to $\varphi^2(a) = a$. Of these only the 4-cycles {5,65,45,25} and {15,35,55,75} are also subalgebras. We conclude that the oval of a Type IV cubic contains only two 4-cycles which are also subalgebras. We have proved most of the next result.

Figure 43: idempotent elements and 2-cycles on oval

Theorem 49 *A Type IV cubic curve γ contains exactly four 4-cycles which are also subalgebras of (γ,δ). These occur two on the oval and two on the bell of γ, and all together they form a subalgebra of (γ,δ) equivalent to a $(Z_2 \times Z_8, (0,2))$ circle system.*

Proof The two 4-cycle subalgebras on the oval we found in a $(Z_{80}, 0)$ circle system can also be found in a $(Z_{16}, 0)$ circle system. These are {1,13,9,5} and {3,7,11,15} as illustrated in Figure 44. If Z is the center of a circle ω which inverts λ onto itself, then the mapping η: $\gamma \rightarrow \gamma$ defined by $\eta(a) = Z*a$ is an automorphism of (γ,δ): it is easily seen to be a bijection, and $\delta(\eta(a), \eta(b), \eta(b)) = \delta(Z*a, Z*b, Z*c) = Z*\{(Z*(Z*a))*((Z*b)*(Z*c))\} = Z*\{a * (\infty*(b*c))\} = (Z*\infty)*\{a * (\infty*(b*c))\} = (Z*a)*(b*c) = Z* \{Z* ((Z*a)*(b*c))\} = Z * \delta(a,b,c) = \eta(\delta(a,b,c))$ shows that η is a homomorphism. This is also clear if we consider that a, Z*a are inverses with respect to ω, and inversion maps circles to circles. The images of the subalgebras {1,13,9,5} and {3,7,11,15} under η are cyclic subalgebras of the bell. Since {1,13,9,5} and {3,7,11,15} are the only 4-cycles which are subalgebras of the oval, we can conclude that {Z*1, Z*13, Z*9, Z*5} and {Z*3, Z*7, Z*11, Z*15} are the only subalgebraic 4-cycles on the bell of γ.

To see that the 16 points Q ={1,3,5,7,9,11,13,15, Z*1, Z*3, Z*5, Z*7, Z*9, Z*11, Z*13, Z*15} form a subalgebra of (γ,δ) observe that the points S = {1,3,5,7,9,11,13,15} form the intersection of two subalgebras of (γ,δ), namely the $(Z_{16}, 0)$ circle system and the oval, hence they form a subalgebra. Since η is a homomorphism, the points Z*S also form a subalgebra of (γ,δ). Thus if a,b,c $\in$Q all lie on the bell, then δ(a,b,c) $\in$ Q, and if a,b,c all lie on the oval, then δ(a,b,c) $\in$ Q. The remaining possibility is that a,b lie on the oval and Z*c lies on the bell. In this case we must show that δ(Z*c, a, b) $\in$ Q. By Lemma 30, the points of the $(Z_{16}, 0)$ circle system Ω form a perfect 16-gon and thus if p,q, s,t $\in \Omega$ and p+q = s+t then p*q = s*t. Since the points a,b,c come from Ω we have a*b = c*d if d = a+b-c, and thus δ(Z*c, a, b) = Z*{(Z*(Z*c))*(a*b)} = Z*{c*(c*d)} = Z*d $\in$ Q. This completes the proof that Q is a subalgebra of (γ,δ). Finally, the relabeling of the points of Q given by 2k-1→(1,k) and Z*(2k-1)→(0,-k+1) as illustrated in Figure 43, produces a $(Z_2 \times Z_8, (0,2))$ circle system. □

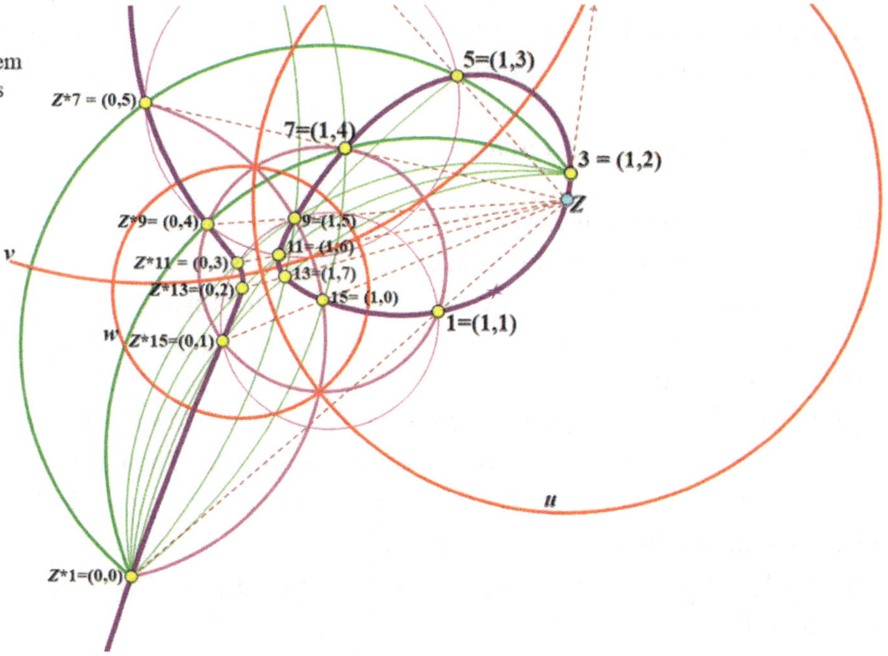

Figure 44: $(Z_2 x Z_8, (0,2))$ circle system formed by four cyclic subalgebras

As we observed at the beginning of this Section, a $(Z_2 \times Z_8, (1,2))$ circle system Ω consists four cyclic subalgebras of order 4. By Theorem 49, a self-inversive cubic curve contains exactly four such subalgebras. Thus, if Ω lies on a self-inversive cubic then it must consist of all four of these subalgebras. But then, in accordance with Theorem 49, Ω must be equivalent to a $(Z_2 \times Z_8, (0,2))$ circle system, which contradicts the result of Theorem 43. We conclude that a $(Z_2 \times Z_8, (1,2))$ circle system cannot exist on a self-inversive cubic curve.

Now suppose Ω is a $(Z_2 \times Z_{8q}, (1,2))$ circle system for some positive integer q, and consider the following partition of the vertex set of Ω. Let Q = {0,1,2, …, 2q-1}.

$$A = \{(0,4k+2) : k \in Q\} \cup \{(1,4k) : k \in Q\}$$

$$B = \{(1,4k+2) : k \in Q\} \cup \{(0,4k) : k \in Q\}$$

$$C = \{(0,4k+1) : k \in Q\} \cup \{(1,4k+3) : k \in Q\}$$

$$D = \{(1,4k+1) : k \in Q\} \cup \{(0,4k+3) : k \in Q\}.$$

Each of these subsets is easily seen to be a subalgebra of (Ω,δ). Now let Λ_t denote a $(Z_{4t},1)$ circle system on γ, then the mapping ψ: $A \rightarrow \Lambda_t$ defined by $\psi(s,t) = \frac{t}{2}$ is an isomorphism: $\psi(\delta((s_1,t_1), (s_2,t_2), (s_3,t_3)) = \psi(1- s_1 - s_2 - s_3, 2 - (t_1 + t_2 + t_3)) = 1 - \frac{t1+t2+t3}{2}$ and $\delta(\psi(s_1,t_1), \psi(s_2,t_2), \psi(s_3,t_3)) = \delta(\frac{t}{2}, \frac{t}{2}, \frac{t}{2}) = 1 - \frac{t1+t2+t3}{2}$. The same mapping gives an isomorphism between B and Λ_t. The mapping ϕ: $C \rightarrow \Lambda_t$ defined by $\psi(s,t) = \frac{-t+1}{2}$ is also easily seen to be an isomorphism, the same mapping also providing an isomorphism between D and Λ_t. We have shown that *a $(Z_2 \times Z_{8q}, (1,2))$ circle system can be partitioned into four subalgebras each isomorphic to a $(Z_{4q},1)$ circle system.* We will show that these comprise *all* the $(Z_{4q},1)$ circle systems that lie on γ.

Let Λ be a $(Z_{4q},1)$ circle system and suppose Λ lies on the oval of a Type IV self-inversive cubic curve γ. Let X be a point on the bell of γ which is the center of a circle which inverts γ onto itself. First we will show that Λ can be extended to a $(Z_{8q},2)$ circle system on the oval. Then we will further show that the $(Z_{8q},2)$ circle system can be extended to a $(Z_{16q}, 0)$ circle system on the oval. The uniqueness of this last system will allow us to conclude that there exist exactly two $(Z_{4q},1)$ circle systems on the oval.

Claim: $\Lambda \cup (X^*\Lambda)$ is a $(Z_{8q},2)$ circle system.

Proof Relabel the points of $\Lambda \cup (X^*\Lambda)$ by the rule $t \leftarrow 2t$ and $X^*t \leftarrow -2t+1$ and let Γ denote the resulting collection of relabeled points. Let a, b, c, d $\in \Gamma$ with a+b+c+d = 2 (mod 8q). To show that (a,b,c,d) is a circle we consider three cases: (i) suppose a,b,c,d are all even with $a = 2a_0$, $b = 2b_0$, $c = 2c_0$, $d = 2d_0$ where a_0, b_0, c_0, $d_0 \in \Lambda$. Then a+b+c+d = 2 implies $a_0 + b_0 + c_0 + d_0 = 1$, so (a_0, b_0, c_0, d_0) is a circle of Λ and thus a circle in Γ. (ii) suppose a,b,c,d are all odd with $a = X^* a_0$, $b = X^* b_0$, $c = X^* c_0$, $d = X^* d_0$. Then a+b+c+d = 2 implies $(-2a_0 + 1) + (-2b_0 + 1) + (-2c_0 + 1) + (-2d_0 + 1) = 2$ which yields $a_0 + b_0 + c_0 + d_0 = 1$ again, and thus (a,b,c,d) is a circle in Γ. (iii) Suppose a, b are even and c, d are odd. Then $a = 2a_0$, $b = 2b_0$, $c = -2c_0 + 1$ and $d = -2d_0 + 1$, and $2 = 2a_0 + 2b_0 + (-2c_0 + 1) + (-2d_0 + 1)$ implies $a_0 + b_0 = c_0 + d_0$. Now by Theorem 34, Λ is a perfect polygon and thus $a_0{*}b_0 = X_{a_0+b_0} = c_0{*}d_0$ and $a{*}b = a_0{*}b_0 = c_0{*}d_0 = c{*}d$. Thus $\delta(c, a, b) = X^*\{((X^*c)^*(a^*b)\} = X^* \{(X^*(X^*c_0)^*(a_0{*}b_0)\} = X^*\{c_0{*}(a_0{*}b_0)\} = X^*\{c_0{*}(c_0{*}d_0)\} = X^*d_0 = d$, from which we obtain that (a,b,c,d) is a circle in Γ. □

Note that since $\phi_X : \gamma \rightarrow \gamma$ defined by $\phi_X(a) = X^*a$ is an automorphism of (γ,δ), our $(Z_{8q},2)$ circle system consists of two copies of a $(Z_{4q},1)$ circle system, namely Λ, and $X^*\Lambda$, both of which lie on the oval. We will now employ a *doubling process* to further extend the $(Z_{8q},2)$ circle system. To simplify the doubling process we first argue that the vertices of the $(Z_{8q},2)$ system can be relabeled so they occur in increasing order as they are encountered in a clockwise traversal of the oval.

By Theorem 34, the vertices of our $(Z_{8q},2)$ circle system form a perfect polygon P on the oval. There is a constant *increment* ε between consecutive vertices of any perfect polygon. If n is the order of the perfect polygon, then ε can be any value in U_n, the group of elements in Z_n with multiplicative inverses. Thus the points of P occur in the order $\{0, \varepsilon, 2\varepsilon, 3\varepsilon, \ldots, (8q\text{-}1)\varepsilon\}$ as the oval is traversed in a clockwise direction. The relabeling $k\varepsilon \rightarrow k$ yields a perfect polygon with vertices in clockwise order $\{0,1,2, \ldots, 8q\text{-}1\}$. With this relabeling, P continues to be a $(Z_{8q},2)$ circle system, for if $(k_1\varepsilon, k_2\varepsilon, k_3\varepsilon, k_4\varepsilon)$ is a circle of Γ, then $k_1\varepsilon+k_2\varepsilon+k_3\varepsilon+k_4\varepsilon = 2$ implies $\varepsilon(k_1+k_2+k_3+k_4) = 2$, so our relabeling yields a $(Z_{8q}, 2\varepsilon^{-1})$ circle system where gcd(8q, ε^{-1}) = 1. Multiplying each point of this circle system by ε then yields an equivalent (Z_{8q},2) circle system. So Γ with its vertices relabeled in this fashion remains a (Z_{8q},2) circle system.

With this version of Γ we now double the labels on all vertices and insert odd points as follows. If 2t, 2t+2 are consecutive even points, choose point 2t+1 on the arc between points 2t, 2t+2 so that (2t+1)*(2t+1) = 2t(2t+2) = X_{4t+2}. This *doubling* process produces a (Z_{16q}, 4) circle system θ, as we demonstrate in the next Theorem.

Theorem 50 *Let P be a perfect n-gon with vertices in consecutive order around the oval of a Type IV cubic curve γ which inverts onto itself via a circle with center X on the bell. Double the label on each point of P and assign odd vertices as follows. If t, t+2 are two consecutive points of P, choose vertex t+1 on the arc of the oval between t, t+2 so that (t+1)*(t+1) = t*(t+2). Let P' denote the collection of points so labeled. Then P' is a perfect 2n-gon. If the vertices of P form a (Z_n, g) circle system with X_g = X then the vertices of P' form a (Z_{2n}, 2g) circle system.*

Proof Suppose a,b,c,d are vertices of P' with a+b = c+d. In all cases we must show that a*b = c*d. If a,b,c,d are all even, then we have a*b = c*d as in P. Now suppose e, f are odd points, then e*e = (e-1)*(e+1) and f*f = (f-1)*(f+1). Since e-1, e+1, f-1, f+1 are even and (e+1)+(f-1) = (e-1)+(f+1) we have (e+1)*(f-1) = X_{e+f} = (e-1)*(f+1). Consider: (X_{e+f} * e)* (X_{e+f} * e) = (X_{e+f} * X_{e+f})* (e*e) = (X_{e+f} * X_{e+f})* ((e-1)*(e+1)) = (X_{e+f} *(e-1))* (X_{e+f} *(e+1)) = ($X_{e-1+f+1}$ * (e-1))*($X_{e+1+f-1}$ * (e+1)) = (f+1)*(f-1) = f*f. Now since both f and X_{e+f} * e lie between f+1 and f-1 we must have X_{e+f} * e = f. If a, b are even and a+b = e+f, then e*f = X_{e+f} = (e-1)*(f+1) = a*b. If g, h are odd with g+h = e+f, then e*f = a*b = (g-1)*(h+1) = X_{g+h} = g*h. Finally suppose a is even, b is odd, c is even and d is odd with a+b = c+d. Since a*(a+2) = (a+1)*(a+1) we have a = (a+2)*{(a+1)*(a+1)}, and b = (a+1)* X_{a+b+1}. Then a*b = {(a+2)* {(a+1)*(a+1)}}*{(a+1)* X_{a+b+1}} = {(a+2)* {(a+1)*(a+1)}}*{(a+1)* {(a+2)*(b-1)}} = (a+1)*(b-1). From this it follows that a*b = (a+t)*(b-t) for any t ∈ Z_{2n}. Letting t = c-a we obtain a*b = c*(a+b – c) = c*d. We have shown that P' is a perfect 2n-gon. Now suppose P is a (Z_n, g) circle system and a,b,c,d ∈ P' with a+b+c+d = 2g. We must show that (a,b,c,d) is a circle. Consider δ(a,b,c) = X*{(X*a)*(b*c)} = X_g * {(X_g*a)*X_{b+c}} = X_g * {(g-a)*X_{b+c}} = X_g * (a+b+c – g) = g - (a+b+c – g) = 2g – (a+b+c) = d. By Theorem 16 the points {a,b,c,d} are cocyclic. □

Recall that in our construction of Γ, the points {X, 2t, -2t+1} are collinear. Thus in the perfect polygon corresponding to Γ, we have X_1 = X, then after doubling we have X_2 = X in θ. By Theorem 50, the doubling of Γ yields a (Z_{16q}, 4) circle system. We have shown that a (Z_{4q},1) circle system on the oval of a Type IV cubic can be extended to a (Z_{16q}, 4) circle system on the oval.

Now the key point is that, up to equivalence, there is a unique (Z_{16q}, 4) circle system on the oval. To justify this statement note that by subtracting 1 from each point of our (Z_{16q}, 4) circle system we obtain an equivalent (Z_{16q}, 0) circle system Σ. In this system the points {0, 4q, 8q, 12q} are δ – idempotent and must occupy the four points of intersection with the oval of the two circles with centers X = X_0, Y = X_{8q} on the bell which invert γ onto itself. No matter how these points are placed, the positions of the remaining vertices are forced. The circle system Σ also has a partition into four subalgebras with Q = {0,1,2, …, 4q-1}:

$$A = \{4k+1 : k \in Q\};\ B = \{4k+3 : k \in Q\};$$

$$C = \{4k : k \in Q\};\ D = \{4k+2 : k \in Q\}.$$

The mapping ψ: A $\rightarrow$ (Z_{4q}, 1) given by ψ(4k+1) = -k (mod 4q) is an isomorphism, as is the mapping ϕ: B $\rightarrow$ (Z_{4q}, 1) given by ψ(4k+3) = k+1 (mod 4q). Also the mappings ξ: C$\rightarrow$ (Z_{4q}, 0) given by ψ(4k) = k (mod 4q), and η: D $\rightarrow$ (Z_{4q}, -2) given by ψ(4k+2) = k (mod 4q) are isomorphisms. Adding 1 to each vertex of a (Z_{4q}, -2) circle system yields an equivalent (Z_{4q}, 2) circle system, so we see that Σ contains exactly two (Z_{4q}, 1) circle systems. Since we have shown that an arbitrary (Z_{4q}, 1) circle extends to Σ, we can conclude, by the uniqueness of Σ, that there are exactly two (Z_{4q}, 1) circle systems on the oval of a Type IV self-inversive cubic curve.

Recall that our supposed ($Z_2 \times Z_{8q}$,(1,2)) circle system Ω consists of four (Z_{4q}, 1) circle systems, and by what we have just shown two of these must lie on the oval and the other two must lie on the bell of a Type IV cubic curve γ. Let u denote the unique circle with center on the oval which inverts γ onto itself and let Z denote the center of this circle which must lie on the oval. The mapping ϕ_Z: $\gamma \rightarrow \gamma$ defined by ϕ_Z(a),= Z*a is an automorphism of (γ,δ), and thus if A, B are the two (Z_{4q}, 1) circle systems on the oval, then Z*A, Z*B are the two (Z_{4q}, 1) circle systems on the bell. So it must be that the vertex set of Ω is given by A ∪ B ∪ (Z*A) ∪ (Z*B). The points in A∪B constitute the full set of odd vertices of Σ and must form a subalgebra of Σ. The mapping μ : A∪B $\rightarrow$ (Z_{8q}, 2) defined by μ(2k-1) = k is an isomorphism, so if Y denotes this (Z_{8q}, 2) circle system on the oval then we must have Ω = Y ∪ (Z*Y).

Let K represent the set of points Y ∪ (Z*Y). We will show that K is a ($Z_2 \times Z_{8q}$, (0,2)), and not the intended ($Z_2 \times Z_{8q}$, (1,2)) circle system. Define a mapping ν : K $\rightarrow$ ($Z_2 \times Z_{8q}$, (0,2)) by ν(t) = (0,t) and ν(Z*t) = (1,1-t) for t ∈ Y. Suppose a,b,c are points of K and consider the following cases. (i) *a,b,c* ∈ *Y.* Then ν(δ(a,b,c)) = ν(2–a–b–c) = (0, 2–a–b–c) = (0,2) – (0,a) – (0,b) – (0,c) = δ(ν(a), ν(b), ν(c)). (ii) *Z*a, Z*b, Z*c* ∈ *(Z*Y).* Since ϕ_Z : $\gamma \rightarrow \gamma$ defined by ϕ_Z(t) = Z*t is an automorphism of (γ,δ) we have ν(δ(Z*a, Z*b, Z*c)) = ν(Z* δ(a,b,c)) = (1, 1- δ(a,b,c)) = (1, 1 – (2-a-b-c)) = (1, a+b+c-1). On the other hand δ(ν(Z*a), ν(Z*b), ν(Z*c)) = δ((1,1-a), (1,1-b), (1,1-c)) = (0,2) – {(1,1-a)+ (1,1-b)+(1,1-c)} = (1, a+b+c-1). (iii) *a,b* ∈ *γ, Z*c* ∈ *Z*γ.* Then ν(δ(Z*c, a, b)) = ν(Z*{(Z*(Z*c))*(a*b)) = ν(Z*(c*(a*b)) = ν(Z*(c*X_{a+b}) = ν(Z*(a+ b – c)) = (1, 1+c-a-b), and on the other hand δ(ν(Z*c), ν(b), ν(c)) = δ((1,1-c), (0,b), (0,c)) = (0,2) – (1,1-c) – (0,b) – (0,a) = (1, 1+c-a-b). We have shown that ν is a homomorphism, and since clearly ν is onto, ν is an isomorphism. Thus the points of K form a ($Z_2 \times Z_{8q}$, (0,2)) circle system as illustrated in Figure 45 for q = 2.

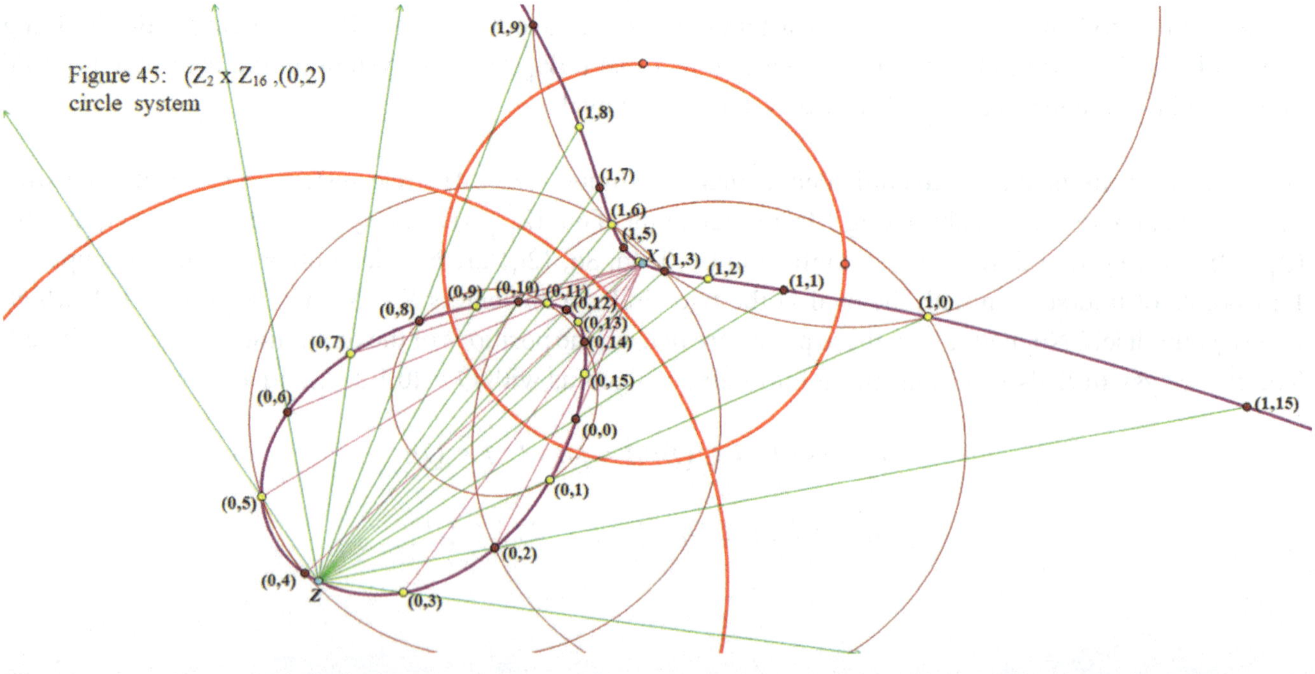

Figure 45: ($Z_2 \times Z_{16}$,(0,2) circle system

Now the point is that by Theorem 43 the $(Z_2 \times Z_{8q}, (0,2))$ and the $(Z_2 \times Z_{8q}, (1,2))$ circle systems are not equivalent, so we can conclude:

Theorem 51 *A $(Z_2 \times Z_{8q}, (1,2))$ circle system cannot be constructed on a self-inversive cubic curve for any positive integer q.* □

And since we feel that all circle systems with more than 8 vertices must lie on a self-inversive cubic we make the following Conjecture.

Conjecture 52 *A $(Z_2 \times Z_{8q}, (1,2))$ circle system does not exist for any positive integer q.*

6.5 Circle systems with base $Z_2 \times Z_2 \times Z_2$

The group $Z_2 \times Z_2 \times Z_2$ is the last type of finite group which we know serves as the base for a circle system. First we determine equivalence classes.

Theorem 52 *Every circle system with base $Z_2 \times Z_2 \times Z_2$ is equivalent to a $(Z_2 \times Z_2 \times Z_2, (0,0,0))$ or a $(Z_2 \times Z_2 \times Z_2, (0,0,1))$ system.*

Proof Let Ω be a $(Z_2 \times Z_2 \times Z_2, (1,1,0))$ circle system and let Γ denote a $(Z_2 \times Z_2 \times Z_2, (0,0,1))$ circle system. Let A = {(0,0,0),(1,1,1),(1,0,1),(0,1,0)} and let B = {(1,1,0),(0,0,1),(0,1,1),(1,0,0)} so $A \cup B = Z_2 \times Z_2 \times Z_2$. Define a mapping $\psi: \Omega \to \Gamma$ by $\psi(a) = a$ if $a \in A$, and $\psi(b) = b + (1,1,1)$ for $b \in B$. Suppose $p,q,r,s \in \Omega$ and $p+q+r+s = (1,1,0)$. This sum cannot be achieved as the sum of two elements from A and two from B, nor can it be obtained as the sum of all four elements from A or all four elements from B. We must have either (i) $p,q,r \in A$ and $s \in B$, or (ii) $p,q,r \in B$ and $s \in A$. In case (i) we have $\psi(p)+\psi(q)+\psi(r)+\psi(s) = p+q+r+s+(1,1,1) = (1,1,0) + (1,1,1) = (0,0,1)$, and in case (ii) we have $\psi(p)+\psi(q)+\psi(r)+\psi(s) = (1,1,1)+p+(1,1,1)+q+(1,1,1)+r+s = p+q+r+s+(1,1,1) = (1,1,0)+(1,1,1) = (0,0,1)$. The mapping ψ is thus an equivalence. Now let Λ denote a $(Z_2 \times Z_2 \times Z_2, (1,1,1))$ circle system and let C = {(0,0,0), (1,1,0),(1,0,0), (0,1,0)} and D = {(0,0,1),(1,1,1),(1,0,1),(0,1,1)} so $C \cup D = Z_2 \times Z_2 \times Z_2$. Define a mapping $\phi: \Gamma \to \Lambda$ by $\phi(c) = c$ if $c \in C$, and $\phi(d) = d+(1,1,0)$ if $d \in D$. Let $p,q,r,s \in Z_2 \times Z_2 \times Z_2$ and suppose $p+q+r+s = (0,0,1)$. Again we find only the two cases: (i) $p,q,r \in C$ and $s \in D$ or (ii) $p,q,r \in D$ and $s \in C$, and similarly ϕ is an equivalence. Since also interchanging coordinates is an equivalence, we obtain the stipulated result. □

In Figure 46 we illustrate a $(Z_2{\times}Z_2{\times}Z_2, (1,1,0))$ circle system.

Figure 46: $(Z_2xZ_2xZ_2, (1,1,0))$ circle system

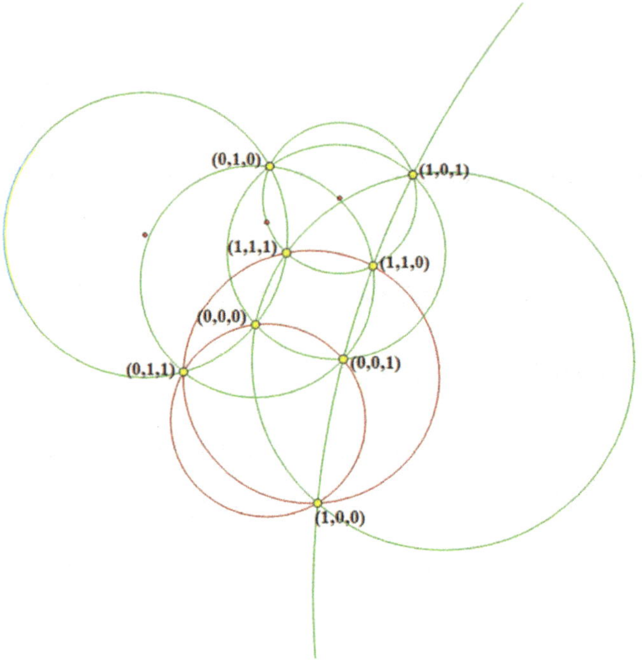

Note that, except for the labeling, the diagrams in Figures 38, 46 are identical. In fact we obtain the $(Z_2{\times}Z_2{\times}Z_2, (1,1,0))$ circle system by keeping the first component of the points in the $(Z_2{\times}Z_4, (1,0))$ system and then replacing the second component by its binary equivalent. One would expect the two systems to be equivalent, but this is not the case. Suppose $\nu: \Gamma \to (Z_2{\times}Z_4, (1,0))$ is an equivalence. The points (0,0,0), (0,0,1) lie on no common circle in Γ, so ν(0,0,0), ν(0,0,1) must lie on no common circle in the $(Z_2{\times}Z_4, (1,0))$ system. In the latter system only the pairs {(0,0),(1,2)}, {(0,1),(1,3)}, {(0,3),(1,1)}, {(1,0),(0,2)} lie on no common circle, so {ν(0,0,0), ν(0,0,1)} must be one of these pairs. Suppose ν(0,0,0) = x and ν(0,0,1) = y, then since (0,0,0) + (0,0,0)+(0,0,0)+(0,0,1) = (0,0,1), we must have 3x+y = (1,0). But for each of our pairs 3x+y = (1,2). We conclude that no equivalence between Γ and a $(Z_2{\times}Z_4, (1,0))$ circle system is possible. Similar to the $(Z_2{\times}Z_4, (1,0))$ circle system, Γ cannot be constructed on an irreducible cubic curve : if x is any element of Γ, then δ((0,0,0), x, x) = (0,0,1), which implies that x*x has the same value for all eight choices for x. This is not possible since on any irreducible cubic there can be at most four points with the same square. The existence of a $(Z_2{\times}Z_2{\times}Z_2, (0,0,0))$ is doubtful since there are too many circles on too few points.

Chapter 7

Subalgebras of Circle Systems with Base Z_n

7.1 The φ-graph

In this Chapter we determine the subalgebra structure of a circle system which is justified on a self-inversive cubic curve γ. To this end it is natural to consider the subalgebras generated by a single point a of γ. Certainly the cycle $\{a, \varphi(a), \varphi^2(a), \ldots\}$ must be contained in such a subalgebra. Let γ be an irreducible self-inversive cubic curve and let (γ,δ) be the THA defined on γ. The mapping $\varphi:\gamma\rightarrow\gamma$ by $\varphi(x) = \delta(x,x,x)$ was introduced in Section 6.4. If $a,b,c \in \gamma$, then $\varphi(\delta(a,b,c)) = \delta(\delta(a,b,c), \delta(a,b,c), \delta(a,b,c)) = \delta(\delta(a,a,a), \delta(b,b,b), \delta(c,c,c)) = \delta(\varphi(a), \varphi(b), \varphi(c))$, and thus φ is an endomorphism of (γ,δ). In Section 6.4 it was noted that as a point t makes one circuit of the oval of a Type IV cubic, the point φ(t) makes three circuits in the opposite direction. As a consequence, if b is any point on the oval, then there exist exactly three points x on the oval which satisfy $\varphi(x) = b$. We may say that the endomorphism φ is *three to one.* Another basic result regarding φ is given in the following Theorem.

Theorem 54 *Let b be any point on an irreducible self-inversive cubic curve γ, then there exist exactly three points x on γ which satisfy $\varphi(x) = b$. Moreover these three points lie on a circle with b.*

Proof Suppose b lies on the oval of a Type IV cubic λ. We have argued above that there exist exactly three points p,q,r on the oval such that $\varphi(p) = \varphi(q) = \varphi(r) = b$. Consider: $\varphi(\delta(p,q,b)) = \delta(\delta(p,q,b), \delta(p,q,b), \delta(p,q,b)) = \delta(\delta(p,p,p), \delta(q,q,q), \delta(b,b,b)) = \delta(b,b, \delta(b,b,b)) = b$, and similarly $\varphi(\delta(p,r,b)) = \varphi(\delta(q,r,b)) = b$. As a consequence $\varphi^{-1}(b)$ contains the points $\{p, q, r, \delta(p,q,b), \delta(p,r,b), (\delta(q,r,b)\}$. Since $\varphi^{-1}(b)$ contains only three points, it must be that $\delta(p,q,b)$ is equal to one of p, q, r. If $\delta(p,q,b) = p$, then we obtain $\delta(p,p,b) = q$, but $\varphi(p) = b$ implies $\delta(p,p,b) = p$. A similar contradiction occurs if we suppose $\delta(p,q,b) = q$, so we must have $\delta(p,q,b) = r$, from which it immediately follows that $\delta(p,q,r) = b$. Now suppose b lies on the bell of λ and let Z denote the point on the oval which serves as the center of a circle v which inverts λ onto itself. The mapping $\eta: \lambda\rightarrow\lambda$ defined by $\eta(x) = Z{*}x$ has been shown in the proof of Theorem 49 to be an automorphism of the THA (λ,δ). Let d ∈ oval with $Z{*}d = b$, and suppose p, q, r are the three points on the oval which satisfy $\varphi(x) = d$. Then $\varphi(Z{*}p)= \varphi\eta(p) = \delta(\eta(p), \eta(p), \eta(p)) = \eta(\varphi(p)) = Z{*}d$, and similarly $\varphi(Z{*}q)= \varphi(Z{*}r)= Z{*}d$. If t ∈ bell and $\varphi(t) = Z{*}d$, then $\varphi(Z{*}t) = \varphi(\eta(t)) = \eta(\varphi(t)) = Z{*} \varphi(t) = d$, so $Z{*}t \in\{p,q,r\}$ and thus $t\in\{Z{*}p, Z{*}q, Z{*}r\}$. So we can conclude that for any point b on the bell of a Type IV cubic, we also have exactly three points, (on the bell) which satisfy $\varphi(x) = b$, and we can conclude from the first part of the proof that these three points lie on a circle with b. If λ is Type V, then it is similar to the bell of a Type IV cubic, so the same result should hold. □

Now we introduce a graph based on the φ function which will be of use in the determination of subalgebras. If S is a finite subalgebra of (γ,δ), the *φ – graph of S, denoted S_φ* is a directed graph with with vertex set S and arc set x⟶φ(x) for x ∈ S. The points of S_φ are either *cyclic,* i.e., belong to a cycle or *acyclic,* i.e., belong to a directed path which terminates at a cyclic point. We will use the structure of S_φ to help determine the structure of S in terms of its subalgebras. Since S is finite, each connected component P of S_φ contains a cycle C which we call the *central cycle* of the component. If p⟶a with a ∈ C and p is not an element of C then we call p a *pendant* vertex. In case P contains pendant vertices we call P a *paracycle.* This leads to our next result.

Lemma 55 *Let S be a finite subalgebra of (γ,δ) and let C be a cycle of S_φ which contains a pendant vertex. Then every cyclic point of S_φ contains exactly two pendant vertices.*

Proof Suppose p is the pendant vertex and p⟶a with a ∈ *C* . Let b ∈ *C* with φ(b) = a. By Theorem 54, δ(a,p,b)⟶a. If δ(a,p,b) = p, then p = δ(a,p,p) = b, which is not the case since p is pendant and b is not. If δ(a,p,b) = b, then also we obtain b = δ(a,b,b) = p. thus the points {b, p, δ(a,p,b)} are distinct and constitute the three elements of $\varphi^{-1}(a)$. Then p, δ(a,p,b) are the two pendant vertices incident at point a. Now let c = φ(a) and consider the point δ(p,p,q) where p,q are the two pendant points at a. Then φ(δ(p,p,q)) = δ(δ(p,p,q), δ(p,p,q), δ(p,p,q)) = δ(δ(p,p,p), δ(p,p,p), δ(q,q,q)) = δ(a,a,a) = c, so δ(p,p,q)⟶c. Also if δ(p,p,q) = a, then b = δ(a,p,q) = p, and thus δ(p,p,q) must be pendant at c. Now the first part of the proof assures a second pendant point at c, and by repeating the argument we obtain that every point of *C* has two pendant vertices. Now suppose C_0 is a second cycle of S_φ and w is a point of C_0. We have w = δ(a,a,f) where f = δ(a,a,w) is a cyclic point since both a, w are cyclic points. Let e be the cyclic point with φ(e) = f, and consider the three distinct points δ(b,b,e), δ(b,p,e), δ(b,q,e). Each of these points satisfies φ(x) = w, and so they constitute $\varphi^{-1}(w)$. Since b,e are cyclic points, so is δ(b,b,e), then we can conclude that the points δ(b,p,e), δ(b,q,e) are pendant at w. □

In Figure 47 we illustrate the φ-graph for a (Z_{12},1) circle system.

Figure 47: graph of (Z_{12},1) circle system

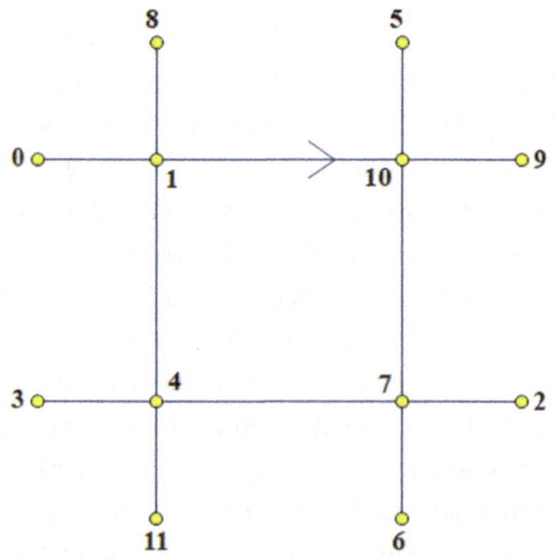

A *complete ternary tree rooted at a point p with depth m* is a directed tree with the property that every directed path terminating at p has length m, and the points on every such path have indegree 3, with the exception of the leafs. We will show that this type of tree is a prominent feature of S_φ.

Lemma 56 *Let S be a finite subalgebra of (γ,δ) and let P be a component of S_φ with central cycle C. Let m denote a maximum length directed path in P which has the form $p_m \to p_{m-1} \to \ldots \to p_1 \to a$ where a is a point of C and p_1 is pendant. Then there is a complete ternary tree of depth m-1 rooted at each pendant point of S_φ.*

Proof Let q denote the second pendant point at a, in accordance with Theorem 54, and define the points $\{v_2, \ldots, v_m\}$ recursively by $v_2 = \delta(p_2, p_2, q)$; $v_k = \delta(p_k, p_k, v_{k-1})$ for $k \geq 3$. Then $\varphi(v_2) = \delta(p_1, p_1, a) = p_1$, and $v_2 \neq p_2$ since $p_2 = \delta(p_2, p_2, q)$ implies $p_1 = \delta(p_2, p_2, p_2) = q$ which is not true. Thus p_2, v_2 are distinct points with $\varphi(p_2) = \varphi(v_2) = p_1$. Now, by Theorem 54, the point $\delta(p_1, p_2, v_2)$ constitutes the third point of S with an arc to p_1. Assume inductively that $v_{k-1} \to p_{k-2}$ with $v_{k-1} \neq p_{k-1}$. Consider $\varphi(v_k) = \varphi(\delta(p_k, p_k, v_{k-1}) = \delta(p_{k-1}, p_{k-1}, p_{k-2}) = pk\text{-}1$ so $v_k \to p_{k-1}$. Also $v_k \neq p_k$ since $p_k = \delta(p_k, p_k, v_{k-1})$ implies $p_{k-1} = \delta(p_k, p_k, p_k) = v_{k-1}$ which is not true. It follows that the points $\{p_k, v_k, \delta(p_k, v_k, p_{k-1})\}$ of S constitute $\varphi^{-1}(p_{k-1})$.

By induction we can conclude that each of the points $p_1, \ldots, p_{m-1}$ has indegree 3 in S_φ. Since this holds for every directed path terminating at point a, we can deduce that there exists in S_φ a complete ternary tree with depth m-1 rooted at the pendant point p_1. Now let $a_m \to a_{m-1} \to \ldots \to a_1 \to a$ be a directed path consisting of points in the cycle C and define the points $q = q_1 = \delta(a_1, p_1, a)$; $q_k = \delta(a_k, p_k, p_{k-1})$ for $k = 2, \ldots, m$. We then obtain the directed path $q_m \to q_{m-1} \to \ldots \to q_1 \to a$. It follows from the first part of the proof that there is a complete ternary tree in S_φ with depth m-1 rooted at $q = q_1$. Now suppose $w = \varphi(a)$, and let t_1 be one of the two pendant points at w. Since we have shown that there exist complete ternary trees in S_φ rooted at p_1 and q_1, we can find directed paths $q_m \to q_{m-1} \to \ldots \to q_1 \to a$ and $p_m \to p_{m-1} \to \ldots \to p_1 \to a$. Let $y_1 = \delta(p_1, q_1, t_1)$,then $\varphi(y_1) = \delta(a, a, w) = a$, so $y_1 \to a$ and we can find a directed path $y_m \to y_{m-1} \to \ldots \to y_1 \to a$ in S_φ. For $k = 2, \ldots,$ m let $t_k = \delta(p_k, q_k, y_k)$, then $\varphi(t_2) = \varphi(\delta(p_2, q_2, y_2) = \delta(p_1, q_1, y_1) = \delta(p_1, q_1, \delta(p_1, q_1, t_1)) = t_1$. So $t_2 \to t_1$, and by a simple induction we obtain the directed path $t_m \to t_{m-1} \to \ldots \to t_1 \to w$.

By the first part of the proof we obtain a complete ternary tree in S_φ with depth m-1 rooted at t_1, and if s_1 is the second point pendant at w, we obtain a complete ternary tree rooted at s_1. Repeating the argument around the cycle *C* we obtain complete ternary trees with depth m-1 rooted at each pendant point of *C*. Now suppose C_0 is a second cycle in S_φ, and $s_k \to s_{k-1} \to \ldots \to s_1 \to u$ has maximum length among all directed paths which terminate at a cyclic point $u \in C_0$. Then $\varphi^k(s_k) = u$, and in fact, if P_0 is the component of S_φ with central cycle C_0, then $\varphi^k(P_0)$ is a pure cycle with no pendant points. However, if $k<m$, $\varphi^k(P)$ still contains pendant points. Since $\varphi:\lambda \to \lambda$ is an endomorphism of S, $\varphi^k(S)$ is a subalgebra of S. This subalgebra contains the pure cycle $\varphi^k(P_0)$ as well as the component $\varphi^k(P)$ which contains pendant points. This contradicts Lemma 55. The argument is now complete because we can select for P a component of S_φ which contains the longest directed path of the form $p_m \to p_{m-1} \to \ldots \to p_1 \to a$ where a is a cyclic point and p_1 is pendant. We can conclude that $k = m$, and by the first part of the proof, a complete ternary tree with depth m-1 is rooted at each pendant point of C_0. □

In Figure 48, S_φ is illustrated for $S = (Z_{45}, 0)$. From the proof of Lemma 55, it is clear that the component containing point 1 is generated by 1, and its not hard to determine that 1 generates all of S, $(S = \langle 1 \rangle_\delta)$. (If x is an element of a subalgebra S of (λ,δ) then we use the notation $\langle x \rangle_\delta$ to denote the subalgebra of S generated by x; if G is an abelian group which serves as the base of a circle system and $x \in G$, we use the notation $\langle x \rangle$ to denote the subgroup of G generated by x.)

Figure 48: Graph of $(Z_{45},0)$ circle system

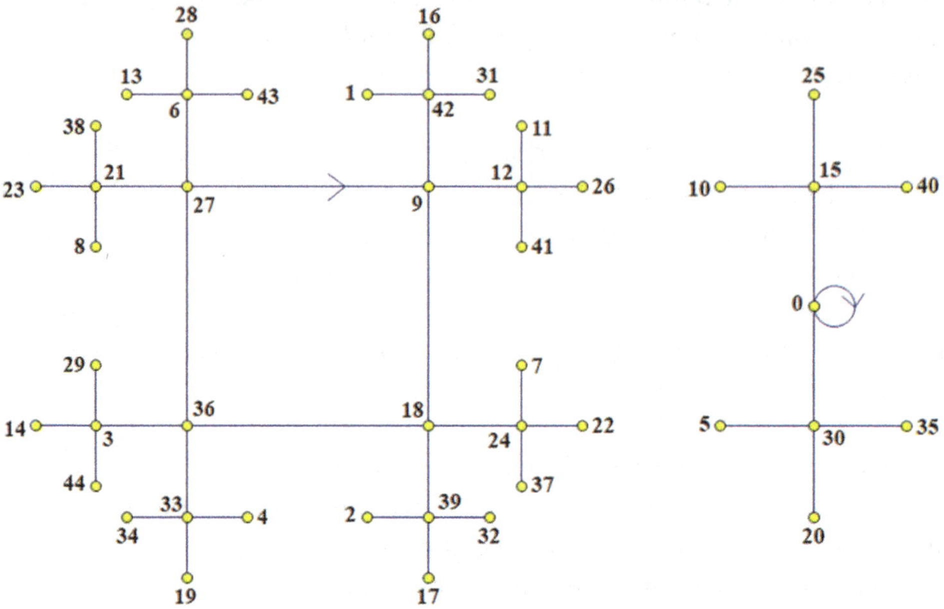

From the symmetry of S_φ we can see that S is generated by any of the leaf nodes in the component containing 1. These leaf nodes constitute precisely those elements of Z_{45} which are relatively prime to 45. From the graph we also see that $\varphi(S) = \langle 3\rangle$ and $\varphi^2(S) = \langle 9\rangle$. Since $\{0\}$ is an idempotent subalgebra, we also obtain the subalgebras $\varphi^{-1}\{0\} = \langle 15\rangle$, and $\varphi^{-2}\{0\} = \langle 5\rangle$. In short, $\langle d\rangle$ is a subalgebra of the $(Z_{45},0)$ circle system iff $d|45$.

7.2 Subalgebras of Circle Systems with Base Z_n, (n odd)

We are now prepared to determine the subalgebra structure of circle systems with base Z_n. The structure changes significantly depending on the parity of n, so in this Section we consider the case n odd, and in Section 7.3 the case n even. Let γ be a Type IV self-inversive cubic curve and let $\Pi_n = \{x \in \text{oval}: \varphi^n(x) = x\}$. If $a,b,c \in \Pi_n$,then $\varphi^n(\delta(a,b,c)) = \delta(\varphi^n(a), \varphi^n(b), \varphi^n(c)) = \delta(a,b,c))$, and thus Π_n is a subalgebra of (γ,δ). A description of this subalgebra is given in the next Theorem.

Theorem 57 *Π_n is equivalent to a $(Z_{3^n-1},0)$ circle system if n is even, and Π_n is equivalent to a $(Z_{3^n+1},0)$ circle system if n is odd.*

Proof As in the discussion at the beginning of Section 6.4, let $\mu_t = |\Pi_t|$. We found there that $\mu_t = 3^t+1$ if t is odd, and $\mu_t = 3^t-1$ if t is even. Let Ω be a $(Z_{3^n+1},0)$ circle system with n odd, and let $b\in\Omega$. Consider the sequence b, $\varphi(b) = -3b$, $\varphi^2(b) = 9b$, …, $\varphi^n(b) = -3^n b$. Since $-3^n b \equiv b \bmod(3^n + 1)$, we have $\varphi^n(b) = b$. Thus every element of Ω satisfies $\varphi^n(x) = x$ with the consequence that Ω is equivalent to Π_n. Now let Γ be a $(Z_{3^n-1},0)$ circle system with n even, let $b\in\Gamma$ and consider the sequence b, $\varphi(b) = -3b$, $\varphi^2(b) = 9b$, …, $\varphi^n(b) = 3^n b$. Since $3^n b \equiv b \bmod(3^n - 1)$, we have $\varphi^n(b) = b$, thus every element of Γ satisfies $\varphi^n(x) = x$ and thus Γ is equivalent to Π_n. □

Now lets say that a subalgebra S is *cyclic* in case S_φ consists only of cycles, (i.e., no pendant points or ternary trees). Suppose S is a finite cyclic subalgebra of the oval of a Type IV cubic curve λ, and let m equal the least common multiple of the orders of the cycles belonging to S_φ. Then $\varphi^m(s) = s$ for any point $s\in S$, and thus S is a subalgebra of Π_m. The mapping $\eta:\lambda\to\lambda$ defined by $\eta(x) = Z^*x$, introduced in Theorem 49, provides an isomorphism between the bell and the oval, so we obtain the same result in case S lies entirely on the bell, (in this case we take $\Pi_n = \{x \in \text{bell}: \varphi^n(x) = x\}$). We now have the following result.

Theorem 58 *Let γ be a Type IV self-inversive cubic curve and let S be a finite cyclic subalgebra which either lies entirely on the bell or entirely on the oval, and let m denote the least common multiple of the cycle lengths in S_φ. If m is odd then S is a subalgebra of a $(Z_{3^m+1},0)$ circle system, and if m is even then S is a subalgebra of a $(Z_{3^m-1},0)$ circle system.* □

The significance of Theorem 58, is that it allows us to determine the finite cyclic subalgebras of the oval, by determining the subalgebra structure of a circle system. If we define $\Lambda_n = \{x \in \text{bell} : \varphi^n(x) = x\}$ then we obtain an analogous result to Theorem 57, just substituting Λ_n for Π_n in the Theorem statement. This is so since the mapping η is an automorphism of (γ,δ) which maps the oval onto the bell and sets up a correspondence between the subalgebras of the oval and the subalgebras of the bell. Since noncyclic subalgebras of (γ,δ) have the form $\varphi^{-m}(S)$ where m is a positive integer and S is a cyclic subalgebra, we should be able to obtain a description of these subalgebras as well. With this in mind our next task is to describe the subalgebra structure of circle systems with base Z_n . The following two Lemmas will be useful. A *word* w in the TH variety is a properly parenthesized expression consisting of any number of variables and the δ operator. The *length* $|w|$ of w is the number of variables, counting repeated variables, which occur in w.

Lemma 59 *Let $\omega(x)$ denote a word in the TH variety which involves only the single variable x. If Ω is a member of the TH variety and $b \in \Omega$, then $\omega(b)$ denotes the member of Ω obtained by substituting b for x in $\omega(x)$. The mapping $\omega(x)$ is an endomorphism of Ω which commutes with φ.*

Proof By induction on $|\omega|$. The general form of $\omega(x)$ is $\omega(x) = \delta(\alpha(x), \beta(x), \gamma(x))$ where α, β, γ are termed the *principal subwords of ω,* each having length less than $|\omega|$. For any word u(x) of short length, (lengths 1, 3, 5, 7 are possible), it is easily seen that u(x) is an endomorphism of Ω. So we may assume inductively that $\alpha(x)$, $\beta(x)$, $\gamma(x)$ are endomorphisms of Ω. Then

$$\omega(\delta(x,y,z)) = \delta[\alpha(\delta(x,y,z)),\ \beta(\delta(x,y,z)),\ \gamma(\delta(x,y,z))]=$$

$$\delta[\delta(\alpha(x),\alpha(y),\alpha(z)),\ \delta(\beta(x),\beta(y),\beta(z)),\ \delta(\gamma(x),\gamma(y),\gamma(z))] =$$

$$\delta[\delta(\alpha(x),\beta(x),\gamma(x)),\ \delta(\alpha(y),\beta(y),\gamma(y)),\ \delta(\alpha(z),\beta(z),\gamma(z))] = \delta[\omega(x),\ \omega(y),\ \omega(z)],$$

and thus $\omega(x)$ is an endomorphism of Ω. It follows immediately that $\omega(\varphi(x)) = \varphi(\omega(x)$. □

We call ω a *word endomorphism.* In the following Lemma the one-generator subalgebras of circle systems with base Z_n are described.

Lemma 60 *Let Ω be a $(Z_n,1)$ circle system, then $\langle 0\rangle_\delta = \Omega$, and if n = 2m is even, then $\langle m\rangle_\delta = \Omega$. Let Γ be a $(Z_n,0)$ circle system. Then $\langle 1\rangle_\delta = Q = \{4t+1: t \in Z_n\}$, and $R = \{4t-1: t \in Z_n\}$ are subalgebras of Γ. If b is any element of Γ, then $\langle b\rangle_\delta = bQ$. If n is odd then $\langle 1\rangle_\delta = \Gamma$. If Λ is a $(Z_n,2)$ circle system and b is any element of Λ, then $\langle b\rangle_\delta = \{-2t + (4t+1)b : t \in Z_n\}$. If n is even, let E denote the even elements of Z_n and let D denote the odd elements. Then $\langle 1\rangle_\delta = D$ and $\langle 0\rangle_\delta = E$. If n is odd then $\langle 1\rangle_\delta = \langle 0\rangle_\delta = \Lambda$.*

Proof Consider: for Ω, $\delta(0,0,0) = 1$ and thus $\delta(0,1,1) = -1$ lies in $\langle 0\rangle_\delta$. Assume inductively that $\{0, \pm 1, \pm 2, \ldots, \pm k\}$ lie in $\langle 0\rangle_\delta$. Then so do $\delta(0,0,-k) = k+1$, and $\delta(1,1,k) = -k-1$. By induction we have $\langle 0\rangle_\delta = \Omega$. Now suppose $n = 2m$ is even, then in Ω we have $\delta(m,m,m) = 1\text{-}3m = m+1$ and $\delta(m, m+1, m+1) = -1\text{-}3m = m\text{-}1$. Thus $\{m\text{-}1, m, m+1\} \subseteq \langle m\rangle_\delta$. Now assume inductively that $\{m\text{-}k, \ldots, m, \ldots, m+k\} \subseteq \langle m\rangle_\delta$ for some positive integer k. Then $\langle m\rangle_\delta$ also contains $\delta(m+1, m+1, m+k) = m-(k+1)$, and $\delta(m\text{-}k, m, m) = m+(k+1)$, so we obtain $\langle m\rangle_\delta = \Omega$ by induction.

It is easily checked that Q, R are subalgebras of Γ. It is another simple induction to show that $Q = \langle 1\rangle_\delta$ and it is then clear that $\langle b\rangle_\delta = bQ$. If n is odd then since gcd (n,4) = 1, there exists $t \in Z_n$ such that $4t = 1 \pmod n$. Then $2 = 4t+1 \in \langle 1\rangle_\delta$, and $\delta(-3,2,4k+1) = -4k$ implies that $\{4s : s \in Z_n\} \subseteq \langle 1\rangle_\delta$. Then $\delta(-3,4,4k+1) = 4(-k-1)+2$ implies that $\{4s+2 : s \in Z_n\} \subseteq \langle 1\rangle_\delta$. We then have $2, -2 \in \langle 1\rangle_\delta$ and so $\delta(2,-2, 4k+1) = 4(-k-1)+3$ implies $R = \{4s+3 : s \in Z_n\} \subseteq \langle 1\rangle_\delta$. Along with $Q \subseteq \langle 1\rangle_\delta$, It follows that every element of Z_n and thus every point of Γ lies in $\langle 1\rangle_\delta$.

For the circle system Λ, let $U = \{-2t + (4t+1)b : t \in Z_n\}$. Then $\delta(b,b,b) = 2\text{-}3b$ is obtained in U by taking t = -1, and $\delta(b, 2\text{-}3b, 2\text{-}3b) = -2+5b$ is obtained in U by taking t = 1. Thus U contains elements of $\langle b\rangle_\delta$ for $t \in \{-1,0,1\}$. Assume inductively that for each $t \in \{-k, \ldots, 0, \ldots, k\}$, the corresponding points of U have been shown to lie in $\langle b\rangle_\delta$. Then the points $\{2\text{-}3b, 2\text{-}3b, 2k-(4k\text{-}1)b\}$ lie in $\langle b\rangle_\delta$ and thus $\delta(2\text{-}3b, 2\text{-}3b, 2k-(4k\text{-}1)b) = -2(k+1) + (4k+5)b$ lies in $\langle b\rangle_\delta$. This is the element of U obtained by taking t = k+1. By the inductive hypothesis we have $-2k + (4k+3)b \in \langle b\rangle_\delta$ and thus $\delta(b, b, -2k + (4k+3)b) = 2(k+1) - (4k+3)b \in \langle b\rangle_\delta$. This is the element of U obtained for t = -k-1. We have shown that $U \subseteq \langle b\rangle_\delta$. By a routine computation it can be shown that U is a subalgebra of Λ, and since $b \in U$, we have $\langle b\rangle_\delta \subseteq U$ and thus $U = \langle b\rangle_\delta$. Then from the formula for U we obtain $\langle 1\rangle_\delta = D$, the set of odd values in Z_n, and $\langle 0\rangle_\delta = E$, the set of even values in Z_n. Of course if n is odd, there is no bifurcation into even and odd values, so we obtain $\langle 1\rangle_\delta = \langle 0\rangle_\delta = \Lambda$. □

The following correspondence between members of the TH variety and abelian groups will also prove useful.

Theorem 61 *Let (V,δ) be a THA which contains a δ-idempotent element e, and define an addition on V by $a+b = \delta(\delta(a,b,e),e,e)$. Then $(V,+)$ is an abelian group with identity e. If we define a ternary operation η on $(V,+)$ by $\eta(a,b,c) = e - a - b - c$, then (V,η) is a THA identical with (V,δ).*

Proof Clearly + is commutative. Consider: $a+e = \delta(\delta(a,e,e),e,e) = a$, so e is an identity for +. Consider: $a + \delta(a,e,e) = \delta(\delta(a, \delta(a,e,e),e),e,e) = \delta(e,e,e) = e$, thus $\delta(a,e,e) = -a$. Finally, to show associativity consider:

$$a+(b+c) = a + \delta(\delta(b,c,e),e,e) = \delta[\delta\{a, \delta(\delta(b,c,e),e,e), e\}, e, e], \text{ and}$$

$$(a+b)+c = \delta(\delta(a,b,e),e,e) + c = \delta[\delta\{\delta(\delta(a,b,e),e,e), c, e\}, e, e] =$$

$$\delta[\delta\{\delta(\delta(a,b,e),e,e), \delta(\delta(b,c,e),b,e), e\}, e, e] =$$

$$\delta[\delta\{\delta(\delta(b,c,e),e,e), \delta(\delta(a,b,e),b,e), e\}, e, e] = \delta[\delta\{\delta(\delta(b,c,e),e,e), a, e\}, e, e].$$

Thus $(V,+)$ is an abelian group. Consider: $\eta(a,b,c) = e+(-a)+(-b)+(-c) =$

$$(e + \delta(a,e,e)) + (\delta(b,e,e) + \delta(c,e,e)) = \delta(a,e,e) + \delta(\delta(\delta(b,e,e), \delta(c,e,e), e), e, e) =$$

$$\delta(a,e,e) + \delta(\delta(\delta(b,c,e), e, e), e, e) = \delta(a,e,e) + \delta(b,c,e) = \delta(\delta(\delta(a,e,e), \delta(b,c,e),e), e, e) =$$

$$\delta(\delta(\delta(a,b,c), e, e), e, e) = \delta(a,b,c).$$

The operations η,δ are identical, and thus (V,η) and (V,δ) are identical THA's. □

It is also possible to travel between (V,δ) and $(V,+)$ via a binary hypercommutative algebra, (BHA), $(V,*)$, again assuming that V contains a δ-idempotent e. If we define $a*b = \delta(a,b,e)$ we transform (V,δ) into a BHA in which e is now idempotent with respect to *. On the other hand, if we start with $(V,*)$ and define $\delta(a,b,c) = a*((b*e)*(c*e))$ we obtain the original THA (V,δ) . Now starting with $(V,*)$ we define an addition by $a \circledcirc b = e*(a*b)$ to obtain an abelian group $(V,\circledcirc)$ which is easily seen to be identical with $(V,+)$. This last transformation is the well known method for defining a group on the points of an irreducible cubic curve, see [3] or [5] . If we define $a*b = -a - b$ then $(V,*)$ is recovered from $(V,\circledcirc)$.

The correspondence between a THA (V,δ) and an abelian group $(V,+)$ given in Theorem 61, induces a correspondence between subalgebras of (V,δ) which contain a δ-idempotent and subgroups of $(V,+)$. This correspondence can be seen in Figure 49 which depicts the containment poset for the subalgebras of a $(Z_{40},0)$ circle system. The correspondence $a+b = \delta(\delta(a,b,e),e,e)$ yields four sublattices corresponding to the four choices $\{0,10,20,30\}$ for the δ-idempotent e. Each sublattice is isomorphic to the lattice of subgroups of Z_{40} by virtue of our correspondence. The idempotent free subalgebras of $(Z_{40},0)$ depicted in the right portion of Figure 49 constitute the subalgebras of the set $1+\langle 2\rangle$ of odd points. Note that every subalgebra of $(Z_{40},0)$ is either a subgroup of Z_{40} or a coset of a subgroup of Z_{40}. We identify exactly which cosets are subalgebras in the following result.

Theorem 62 *Let S be a proper subalgebra of a THA (Ω,δ) which contains an idempotent element e and let $a \in \Omega\text{-}S$. Then $a+S$ is a subalgebra of Ω iff $\varphi(a) \in a+S$.*

Proof Suppose $\varphi(a) = a+s_0 \in a+S$, and let $s_1, s_2, s_3 \in S$. Consider: $\delta(a+s_1, a+s_2, a+s_3) = \delta(\delta(\delta(a,s_1,e),e,e), \delta(\delta(a,s_2,e),e,e), \delta(\delta(a,s_3,e),e,e)) = \delta(\delta(\delta(a,s_1,e), \delta(a,s_2,e), \delta(a,s_3,e)), e, e) = \delta(\delta(\delta(a,a,a), \delta(s_1,s_2,s_3), e), e, e) = \delta(\delta(\varphi(a), s, e), e, e) = \varphi(a) + s = (a+s_0) + s = a + (s_0 + s) \in a+S$. Here we have set $\delta(s_1,s_2,s_3) = s$ which lies in S since S is a subalgebra. Conversely, if $a + S$ is a subalgebra, then $a = a + e \in a+S$, and so we must certainly have $\varphi(a) \in a+S$. □

Now we are ready to proceed with the study of circle systems with base n where n is odd. We begin by considering the $(Z_p,1)$ system where p is an odd prime.

Theorem 63 *Let Ω be a $(Z_p,1)$ circle system where p is an odd prime. If s is the solution mod p to $4x = 1$ (mod p), then $\{s\}$ is the unique proper subalgebra of Ω. Suppose $p \notin 3$ and q is the least positive integer such that $3^q = 1$ (mod p). If q is even then the points in $\Omega - \{s\}$ lie in $(p-1)/q$ cycles each with length q, and if q is odd then the points of $\Omega - \{s\}$ lie in $(p-1)/2q$ cycles each with length 2q.*

Proof If $p = 3$ then $\{1\}$ is clearly the unique proper subalgebra of $(Z_p,1)$. So suppose $p \notin 3$, and suppose $a \in \Omega$. Since $\gcd(3,p)=1$, we have $\langle 3\rangle = Z_p$ and so there exists $t \in Z_p$ such that $3t = 1-a$. Then $\varphi(t) = \delta(t,t,t) = 1 - 3t = a$, i.e., every element of Ω has a preimage under φ and thus the the φ-graph of Ω consists entirely of cycles. Let C_0 denote the cycle which contains the generator 0, and let B denote any one of the other *large* cycles, (i.e., not the 1-cycle consisting of the idempotent s). Let $b \in B$. Since 0 is a generator, there exists a word endomorphism $\omega: \Omega \to \Omega$ such that $\omega(0) = b$. If $s \in \langle b\rangle_\delta$ then by the correspondence given in Theorem 61, $\langle b\rangle_\delta$ must correspond to a subgroup of Z_p, but since p is prime, Z_p has no nontrivial subgroups and consequently $\langle b\rangle_\delta = \Omega$. If $s \notin \langle b\rangle_\delta$ then since $\omega(s) = s$, $\omega^{-1}(\langle b\rangle_\delta)$ does not contain s. But $0 \in \omega^{-1}(\langle b\rangle_\delta)$ implies that $\langle 0\rangle \subseteq \omega^{-1}(\langle b\rangle_\delta)$ since $\omega^{-1}(\langle b\rangle_\delta)$ is a subalgebra of Ω. By Lemma 60, $\langle 0\rangle = \Omega$, and so $\omega^{-1}(\langle b\rangle_\delta) = \Omega$, contradicting $s \notin \omega^{-1}(\langle b\rangle_\delta)$. We conclude that $\langle b\rangle_\delta = \Omega$, and thus that every element of Ω except for s, is a generator of Ω. It then follows that $\{s\}$ is the only proper subalgebra of Ω.

Let $\varphi^m(b)$ represent a typical element of B. Then $\varphi^m(0) \in C_0$ and $\omega(\varphi^m(0)) = \varphi^m(\omega(0))=\varphi^m(b)$. (We have used the result of Lemma 59 that ω and φ commute.) Thus every point of B has a preimage in C_0 under ω and thus $|B| \leq |C_0|$. But now since $\langle b\rangle_\delta$ contains the generator 0, there exists a word endomorphism u such that u(b) = 0 which implies similarly that every point in C_0 has a preimage in B under u. Thus $|C_0|\leq|B|$, and we can conclude that all the large cycles of Ω have the same length. To determine the common size of these cycles recall the Little Theorem of Fermat which states $a^{p-1} = 1$ (mod p) in case p is prime. By Theorem 6, Ω is equivalent to a $(Z_p,0)$ circle system Γ. We will refer to Γ to determine cycle size. According to Lemma 59, $\Gamma = \langle 1\rangle_\delta$ so if C_1 denotes the cycle containing 1, then $C_1 = \{1, -3, 3^2, -3^3, 3^4, \ldots\}$. If q is even then 3^q lies in C_1 and we have $3^q = 1$ so $|C_1| = q$. If q is odd then 3^q is not an element of C_1 and 3^{2q} is the least positive power of 3 which lies in C_1, and thus $|C_1| = 2q$. Since we have shown that all the large cycles of Ω have the same size, the Theorem follows. □

Figure 49: Poset for subalgebras of $(Z_{40},0)$ circle system

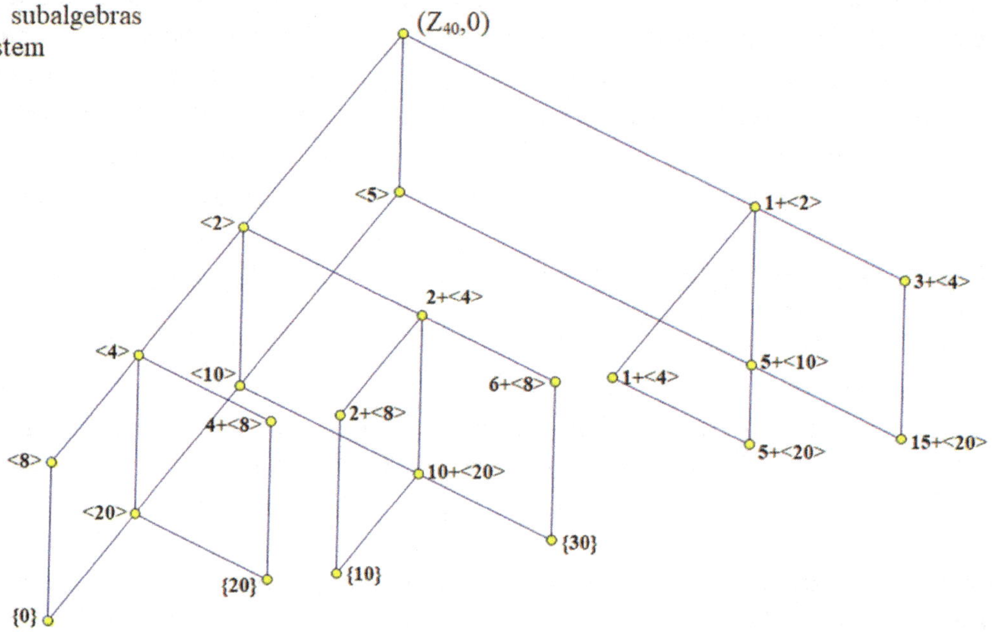

Now we proceed to the study of $(Z_n, 0)$ circle systems where n can be any odd integer.

Theorem 64 *Let Ω be a $(Z_n,0)$ circle system on an irreducible self-inversive cubic curve γ and let (γ,δ) be the THA defined on γ. If n is odd then every subalgebra of Ω is equivalent to a $(Z_d,0)$ circle system for some divisor d of n. Moreover, Ω contains a subalgebra equivalent to $(Z_d,0)$ for each divisor d of n.*

Proof Now if m is any divisor of n, then clearly $\langle m\rangle_\delta$ is a subalgebra of Ω with order $d = \frac{n}{m}$. The relabeling km→k converts $\langle m\rangle_\delta$ into a $(Z_d,0)$ circle system. Thus Ω does indeed contain a $(Z_d,0)$ circle system for every divisor d of n. To show these are the only subalgebras of Ω we will first show that the maximal subalgebras of Ω consist precisely of $\{\langle p\rangle_\delta : p \text{ is prime divisor of } n\}$. Suppose p is a prime divisor of n and b is a point of Ω not in $\langle p\rangle_\delta$. By Lemma 60 we have $\Omega = \langle 1\rangle_\delta$ so there exists a word endomorphism ω of Ω such that ω(1) = b. Then $1 \in \omega^{-1}(\langle b\rangle_\delta)$ and thus $\omega^{-1}(\langle b\rangle_\delta) = \Omega$. This implies that $0 = \omega(0) \in \langle b\rangle_\delta$, and consequently $0 \in \langle b \cup \langle p\rangle\rangle_\delta$. Then by the correspondence given in Theorem 61, $\langle b \cup \langle p\rangle\rangle_\delta$ must be a subgroup of Z_n. Since ⟨p⟩ is a maximal subgroup of Z_n it must be that $\langle b \cup \langle p\rangle\rangle_\delta = \Omega$, and it follows that $\langle p\rangle_\delta$ is a maximal subalgebra of Ω. If we set $d = \frac{n}{p}$, then ⟨p⟩ is equivalent to a $(Z_d,0)$ circle system.

Suppose S is a maximal subalgebra of Ω and S ≠ ⟨p⟩ for any prime divisor p of n. If p is a particular prime divisor of n, then S contains a point b not in ⟨p⟩. But then we have seen that $0 \in \langle b \rangle_\delta$ and thus 0 ∈ S. By the group correspondence S must be a maximal subgroup of Z_n and thus $S = \langle q \rangle_\delta$ for some prime divisor q of n. We obtain a contradiction, so we can conclude that {⟨p⟩ : p is a prime divisor of n} are the only maximal subalgebras of Ω. We may now repeat the argument with the $(Z_d,0)$ circle system in place of Ω, ultimately to obtain that the collection of circle systems $\{(Z_d,0) : d|n\}$ constitute the full set of subalgebras of Ω. □

As a consequence of Theorem 64, we can say, if n is odd, that the subalgebras of a $(Z_n,0)$ circle system form a lattice isomorphic to the lattice of subgroups of Z_n under set containment. By Theorem 6, every circle system with base Z_n, n odd, is equivalent to a $(Z_n,0)$ circle system, so we may now consider the case when n is even.

7.3 Subalgebras of Circle Systems with Base Z_n,(n even)

When n is even we obtain a major distinction in the subalgebra structure of a circle system with base Z_n : The even and odd values form a partition of the circle system into two maximal subalgebras.

Theorem 65 *Let Ω be a $(Z_n,0)$ circle system on an irreducible self-inversive cubic curve γ and let (γ,δ) be the THA defined on γ. If n is even then the subset E of even values in Z_n and the subset D of odd values in Z_n form maximal subalgebras of Ω each equivalent respectively to a $(Z_{n/2},0)$ and a $(Z_{n/2},2)$ circle system.*

Proof It is clear that E, D are subalgebras of Ω. Since E contains the idempotent 0, it corresponds to a subgroup of Z_n by Theorem 61, which must certainly be ⟨2⟩, and so clearly E is maximal. To show that D is maximal, let u ∈ E and consider the subalgebra $\langle D \cup \{u\} \rangle_\delta$ of Ω . Let x ∈ E, then $x = \delta(u, 1-u, -x+1) \in \langle D \cup \{u\} \rangle_\delta$. It follows that $E \subseteq \langle D \cup \{u\} \rangle_\delta$, and thus that $\langle D \cup \{u\} \rangle_\delta = \Omega$. Now the relabeling of E given by 2k→k turns E into a $(Z_{n/2},0)$ circle system, and the relabeling of D given by 2k-1→k yields a $(Z_{n/2},2)$ circle system. □

By Theorem 6, there are two types of circle systems with base Z_n and gcd(n,4) = 2: the $(Z_n,0)$ and $(Z_n,1)$ circle systems. We consider the $(Z_n,1)$ circle system next.

Theorem 66 *Let Ω be a $(Z_n,1)$ circle system on a self-inversive cubic curve γ, with gcd(4,n) = 2. Then every subalgebra of Ω is equivalent to a $(Z_{2d},1)$ circle system where d is an odd divisor of n.*

Proof Note that Ω has no δ-idempotent elements, since φ(a) = a implies 4a = 1 which has no solutions (mod n). On the other hand $\varphi^2(a) = a$ has two solutions: $a = \varphi^2(a) = \varphi(1-3a) = 1 - (3-9a) = 9a-2$ implies 8a = 2 (mod n) which has a solution since gcd(8,n) = 2. In fact there are exactly two solutions; if s is one solution, then s+(n/2) is the second. These two elements form a subalgebra of Ω equivalent to a $(Z_2,1)$ circle system which we denote by ∇. Now let d be an odd divisor of n and define a mapping $\psi_d: \Omega \to (Z_{2d},1)$ by $\psi_d(x) = x \pmod{2d}$. This is a homomorphism, and since ∇ is a subalgebra of the $(Z_{2d},1)$ circle system, $\psi_d^{-1}(\nabla)$ is a subalgebra of Ω. Then the mapping $\xi: \psi_d^{-1}(\nabla) \to (Z_{(n/d)},1)$ defined by ξ(x) = x (mod n/d) is easily seen to be an isomorphism. Since n/d is twice an odd divisor of n, we obtain the subalgebras stipulated in the Theorem statement.

To show these are the only subalgebras of Ω, suppose p is an odd prime divisor of n. We will show first that $\psi_p^{-1}(\nabla)$ is a maximal subalgebra of Ω. Suppose T is a subalgebra of Ω which properly contains $\psi_p^{-1}(\nabla)$ and define a mapping μ: Ω→Γ by μ(x) = x (mod m) where m = n/2 and Γ is a $(Z_m,1)$ circle system. Then μ is a homomorphism and $\mu(\psi_p^{-1}(\nabla)) = s + \langle p \rangle$ where 4s = 1 (mod p). (Note that this implies that $\psi_p^{-1}(\nabla) = s + \langle p \rangle$ where ⟨p⟩ now represents the set of multiples of p in Ω.) In the proof of Theorem 63 we found that the maximal subalgebras of a $(Z_m,0)$ circle system are given by $\{\langle p\rangle$: p is a prime divisor of m$\}$. By Theorem 6, Γ is equivalent to a $(Z_m,0)$ circle system since m is odd. The maximal subalgebra $\langle p\rangle$ of $(Z_m,0)$ corresponds to the subalgebra s + $\langle p\rangle$ of Γ, and thus $\mu(\psi_p^{-1}(\nabla))$ is a maximal subalgebra of Γ. Since $\mu(T)$ is a subalgebra of Γ which properly contains $\mu(\psi_p^{-1}(\nabla))$, it must be that $\mu(T) = \Gamma$. Consider: $\{x \in \Omega : \mu(x) = 0\} = \{0,m\}$, so T must contain one of these elements, but by Lemma 60 each of 0, m is a generator of Ω. Consequently T = Ω, from which we can conclude that $\psi_p^{-1}(\nabla)$ is a maximal subalgebra of Ω.

Next we will show that $\{\psi_p^{-1}(\nabla)$: p is an odd prime divisor of n$\}$ are the only maximal subalgebras of Ω. Suppose S is a maximal subalgebra of Ω. We will argue that $\mu(S)$ is maximal in Γ. Let A be a subalgebra of Γ which properly contains $\mu(S)$ and let $q \in A- \mu(S)$. Since μ is onto Γ, q has a preimage under μ which must lie in $\Omega - S$. Then $\mu^{-1}(A)$ is a subalgebra of Ω which properly contains S. Since S is maximal we obtain $\mu^{-1}(A) = \Omega$. Then $0 = \mu(0)$ lies in A, and since $\langle 0\rangle_\delta = \Omega$, we have $A = \Gamma$, thus showing that $\mu(S)$ is maximal in Γ. As a consequence, μ induces a one-one correspondence between the maximal subalgebras of Ω and the maximal subalgebras of Γ. In accordance with Theorem 64, the maximal subalgebras of Γ are given by $\{s+\langle p\rangle$: p is an odd prime divisor of n, and 4s = 1 (mod p)$\}$, so the maximal subalgebras of Ω are given by $\{\mu^{-1}(s+\langle p\rangle)$: p is an odd prime divisor of n$\}$ = $\{s+\langle p\rangle$: p is an odd prime divisor of n, and 4s = 1 (mod p)$\}$ = $\psi_p^{-1}(\nabla)$. (Although the two set descriptions are optically the same, the sets are not the same; in the first set $\langle p\rangle$ represents all the multiples of p in $Z_{n/2}$, and in the second $\langle p\rangle$ represents all the multiples of p in Z_n.) We have shown that all the maximal subalgebras of Ω are equivalent to $(Z_{n/p},1)$ circle systems. These circle systems can be substituted for Ω in the Theorem statement and the analysis repeated to obtain a full description of the lattice of subalgebras of Ω, which must be as described in the Theorem statement. The mapping μ induces an isomorphism between this lattice and the lattice of subalgebras of Γ. □

In Figure 50 we illustrate the lattice of subalgebras of a $(Z_{210},1)$ circle system.

Figure 50: lattice of subalgebras of $(Z_{210},1)$

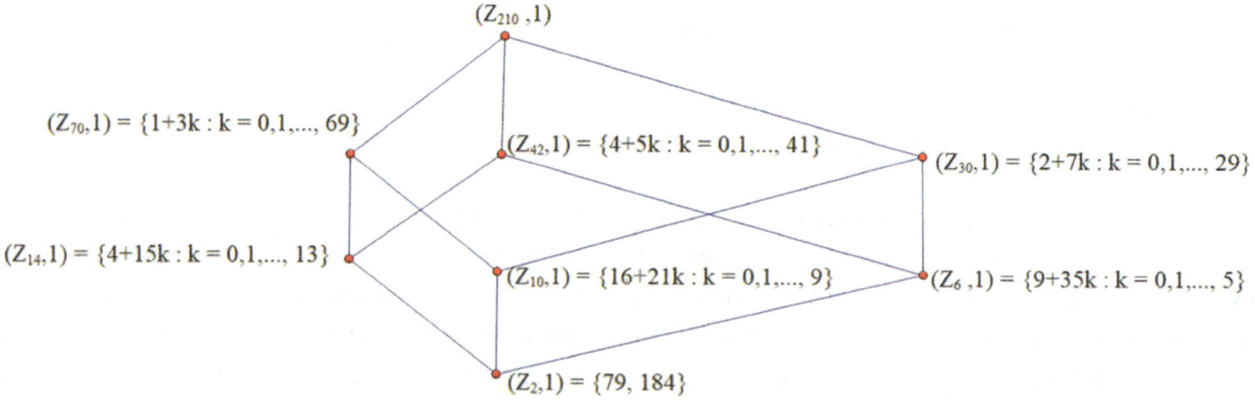

Theorem 66 covers the case in which n = 2d where d is an odd positive integer. Now we consider the more general case in which $n = 2^m d$ where m>1 and d is odd.

Theorem 67 *Let Ω be a $(Z_{2^m d},1)$ circle system on a self-inversive cubic curve γ, with with m,d integers >1 and d odd. If p is a prime divisor of d, then Ω contains a maximal subalgebra W_p = s+ $\langle p\rangle$, 4s = 1 (mod p), equivalent to a $(Z_{\frac{2^m-d}{p}},1)$ circle system and these are the only maximal subalgebras of Ω.*

Proof Let Γ denote a $(Z_{2^m p},1)$ circle system and K denote a $(Z_p,1)$ circle system with p an odd prime. The mapping ξ: Γ→K defined by ξ(x) = x (mod p) is a homomorphism. Since p is odd, K is equivalent to a $(Z_p,0)$ circle system, and thus contains a one element subalgebra we will denote by s. By Theorem 63, {s} is the only proper subalgebra of K, and thus {s} is the unique maximal subalgebra of K. We will show that $\xi^{-1}(s)$ is the unique maximal subalgebra of Γ. Consider: φ(s) = s implies 1-3s = s (mod 2p) or 4s = 1 (mod p). We then have $S = \xi^{-1}(s) = \{s + kp : k = 0,1, \ldots, 2^m-1\} = s + \langle p\rangle$.

To show That S is a maximal subalgebra of Γ suppose T is a subalgebra of Γ which properly contains S. Then there exists a point b in T such that ξ(b) ≠ s, and so ξ(T) is a subalgebra of K which properly contains the maximal subalgebra {s}. As a consequence ξ(T) = K so there exists an element x in T such that ξ(x) = 0. Since $\xi^{-1}\{0\} = \langle p\rangle$ we can say that T contains an element tp for some integer t. Then δ(tp, s – tp, s) = 1-2s lies in T, and thus δ(s,s,1-2s) = 0 lies in T. By Lemma 60, $\langle 0\rangle_\delta = \Gamma$, and consequently T = Γ and S is maximal in Γ.

Next we show that S is the unique maximal subalgebra of Γ. Suppose V is a maximal subalgebra of Γ. Since {s} and K itself are the only subalgebras of K, we must have either ξ(V) = {s} or ξ(V) = K. If ξ(V) = K, we have just seen that this implies V = Γ, but then V is not maximal. So we must have ξ(V) = {s} which yields V = S.

Now define a mapping ψ: Ω→Γ by ψ(x) = x (mod 2^mp) where p is a prime divisor of d. Then since ψ is a homomorphism, $\psi^{-1}(S) = \{s+kp : k = 0,1, \ldots, \frac{2^m d}{p} -1\}$ is a subalgebra of Ω. We will show that $\psi^{-1}(S)$ is a maximal subalgebra of Ω. Suppose W is a subalgebra of Ω which properly contains $\psi^{-1}(S)$. Then there exists an element b in W such that ψ(b) ∉ S. Then ψ(W) properly contains the maximal subalgebra S, and consequently ψ(W) = Γ. Thus there exists x ∈ W such that ψ(x) = 0. Consider: $\psi^{-1}\{0\} = \langle 2^m p\rangle$, and so $2^m pt \in W$ for some integer t. Then $\delta(2^m pt, s - 2^m pt, s) = 1\text{-}2s \in W$, and then δ(s,s,1-2s) = 0 ∈ W, with the result that W = Ω.

Finally, to show that $\psi^{-1}(S)$ is equivalent to a $(Z_{\frac{2^m d}{p}},1)$ circle system, consider: p is an odd prime so p has the form p = 4t+1 or p = 4t-1 for some positive integer t. Suppose p = 4t+1, then 12t + 3 = 0 (mod p) yields 4(3t+1) = 1 (mod p), and so s = 3t+1. Then 1-4s = mp has the solution m = -3.

If p = 4t – 1, then we take s = t, and 1 – 4s = mp has m = -1 as a solution. Define a mapping Φ: $\psi^{-1}(S)\to(Z_{\frac{2^m d}{p}},1)$ by Φ(s+pk) = -k. Then $\Phi(\delta(s+pk_1, s+pk_2, s+pk_3)) = \Phi(1\text{-}3s + p(-k_1 -k_2 -k_3)) = \Phi(s +(1\text{-}4s) + p(-k_1 -k_2 -k_3)) = = \Phi(s +(-p) + p(-k_1 -k_2 -k_3)) = \Phi(s + p(-1 -k_1 -k_2 -k_3)) = 1 + k_1 + k_2 + k_3$, and on the other hand $\delta(\Phi(s+pk_1),\Phi(s+pk_2),\Phi(s+pk_3)) = \delta(-k_1, -k_2, -k_3) = 1 + k_1 + k_2 + k_3$, so Φ is an isomorphism.

Now suppose p has the form p = 4t+1, and define a mapping Ψ: $\psi^{-1}(S)\to(Z_{\frac{2^m d}{p}},1)$ by Ψ(s+pk) = k+1. Consider:

$\Psi(\delta(s+pk_1, s+pk_2, s+pk_3)) = \Psi(s +(1\text{-}4s) + p(-k_1 -k_2 -k_3)) = \Psi(s +(-3p) + p(-k_1 -k_2 -k_3)) = \Psi(s + p(-3 -k_1 -k_2 -k_3)) = -2 -k_1 -k_2 -k_3$, and on the other hand $\delta(\Psi(s+pk_1),\Psi(s+pk_2),\Psi(s+pk_3)) = \delta(k_1+1, k_2+1, k_3+1) = 1 - (3 + k_1 + k_2 + k_3) = -2 -k_1 -k_2 -k_3$, and thus Ψ is an isomorphism. For the sake of clarity, for each odd prime divisor of d, let $\psi_p(x) = x \pmod{2^m p}$ and let S_p denote the unique maximal subalgebra of $\Gamma_p = (Z_{2^m p},1)$. If A is a maximal subalgebra of Ω, then by similar means we obtain that $\psi_p(A)$ is maximal in Γ_p, and thus that $\psi_p(A)=S_p$. Then $A = \psi_p^{-1}(S_p)$, i.e., A is one of the maximal subalgebras of Ω already described. □

Now with regard to $(Z_n,1)$ circle systems (n even), it remains only to determine the subalgebra structure in case $n = 2^m$. For this case we have the following simple result.

Theorem 68 *Let Ω be a $(Z_{2^m},1)$ circle system, then Ω has no proper subalgebras.*

Proof For m = 1, 2 the result is easily checked, so we suppose m≥3. We will show that $\varphi^{2^m}(0) = 0$ and that $\varphi^k(0)$ is not 0 for any positive integer k with $k < 2^m$. It will then follow that the φ- graph of Ω consists of a single cycle, and thus that Ω has no proper subalgebras. Note that $\varphi^k(0) = 0$ implies that k must be even, since the values in any cycle of the φ- graph alternate between odd and even. If k is even then we obtain $4\varphi^k(0) = 1 - 3^k$. We want $\varphi^k(0) = 0$ so we must determine the values of k such that $1 - 3^k = 0 \pmod{2^m}$. We claim that $k = 2^{m-2}$ is one solution, and will prove this claim by induction. It is clearly true for m = 3. Let k ≥3 and assume inductively that $3^{2^{k-2}} - 1 = 0 \pmod{2^k}$. Consider $3^{2^{k-1}} = (3^{2^{k-2}})^2 = (1 + q2^k)^2$ for some integer q, and $(1 + q2^k)^2 = 1 + q2^{k+1} + q^2 2^{2k} = 1 + 2^{k+1}(q + q^2 2^{k-1})$, and thus $3^{2^{k-1}} - 1 = 0 \pmod{2^{k+1}}$. Since $3^{2^{m-2}} = 1 \pmod{2^m}$, we immediately obtain the full solution set (mod 2^m) to the equation $3^x = 1 \pmod{2^m}$, namely $x \in \{2^{m-2}, 2^{m-1}, 3(2^{m-2}), 2^m\}$. Thus $4\varphi^k(0) = 0$ implies $k \in \{2^{m-2}, 2^{m-1}, 3(2^{m-2}), 2^m\}$, and on the other hand $4\varphi^k(0) = 0 \pmod{2^m}$ implies $\varphi^k(0) \in \{2^{m-2}, 2^{m-1}, 3(2^{m-2}), 2^m = 0\}$, so we must have $\{\varphi^{2^{m-2}}(0), \varphi^{2^{m-1}}(0), \varphi^{3(2m-2)}(0), \varphi^{2^{m-2}}(0)\} = \{2^{m-2}, 2^{m-1}, 3(2^{m-2}), 0\}$. Since only one of the values in $\{2^{m-2}, 2^{m-1}, 3(2^{m-2}), 0\}$ is 0, it must be that only one of the values in $\{\varphi^{2^{m-2}}(0), \varphi^{2^{m-1}}(0), \varphi^{3(2m-2)}0), \varphi^{2^m}(0)\}$ is 0. If $\varphi^{2^{m-2}}(0) = 0$, then $\varphi^{2^{m-1}}(0) = \varphi^{2^{m-2}}(\varphi^{2^{m-2}}(0)) = 0$, so we cannot have $\varphi^{2^{m-2}}(0) = 0$. It is similarly shown that $\varphi^{2^{m-1}}(0), \varphi^{3(2m-2)}(0)$ cannot be 0, so we are left with $\varphi^{2^m}(0) = 0$, and the Theorem follows. □

7.4 Subalgebras of $(Z_{2^m d},0)$ and $(Z_{2^m d},2)$ circle systems

In the previous Section we have determined the subalgebra structure of every circle system with base Z_n and sum 1. Also we considered $(Z_n,0)$ circle systems with n odd or with gcd(n,4) = 2. It remains to discuss the subalgebra structure of $(Z_{2^m d},0)$ and $(Z_{2^m d},2)$ circle systems with m≥2. Analogous to Theorem 65, we have the following result concerning $(Z_n,2)$ circle systems.

Theorem 69 *Let Ω be a $(Z_n,2)$ circle system with n even and let E, D denote the even and odd vertices of Ω respectively. Then E, D are maximal subalgebras of Ω both equivalent to a $(Z_{n/2},1)$ circle system.* □

Proof The proof that E, D are maximal is similar to the analogous proof given in Theorem 65, so we omit the details. The mapping $\phi: E \to (Z_{n/2},1)$ given by $\phi(2k) = k$ is an equivalence, as is the mapping $\psi: D \to (Z_{n/2},1)$ defined by $\psi(2k+1) = -k$. □

The correspondence given in Theorem 61 allows us to generalize Theorem 65:

Theorem 70 *Let Ω be a $(Z_{2^m d},0)$ circle system, m ≥1, d odd, and let E, D denote the even and odd vertices of Ω respectively. Then the maximal subalgebras of (Ω,δ) consist of E, D, and {⟨p⟩: p is a prime divisor of d}.*

Proof The sets E,D are maximal by Theorem 69, and since ⟨p⟩ is a maximal subgroup of (Ω,+) for each prime divisor p of d, it follows from the correspondence between subgroups of (Ω,+) and subalgebras of (Ω,δ) given in Theorem 61, that ⟨p⟩ is also maximal in Ω. It remains to show that these are the only maximal subalgebras of Ω.

Let Γ denote a $(Z_d,0)$ circle system and define $\phi: \Omega\to\Gamma$ by $\phi(x) = x \pmod d$. Suppose S is a maximal subalgebra of Ω. Since φ is a homomorphism, φ(S) is a subalgebra of Γ. We will show that either φ(S) = Γ, or φ(S) is a maximal subalgebra of Γ. Suppose A is a subalgebra of Γ which properly contains φ(S). Then $\phi^{-1}(A)$ is a subalgebra of Ω which properly contains S. Since S is maximal, we must have $\phi^{-1}(A) = \Omega$. Then $1 \in \phi^{-1}(A)$, and so $1 = \phi(1) \in A$. By Lemma 60, $\Gamma = \langle 1\rangle_\delta$, and consequently A = Γ, and φ(S) is a maximal subalgebra of Γ. From the proof of Theorem 64 we see that the maximal subalgeras of Γ consist precisely of $\{\langle p\rangle : p \text{ is a prime divisor of } d\}$, thus $\phi(S) = \langle p\rangle$ for some prime divisor p of d, and $S = \phi^{-1}(\langle p\rangle) = d + \langle p\rangle = \langle p\rangle$. (The last equality is due to the fact that $p|d$.)

We have obtained a result in accordance with the Theorem statement, but what about E and D ? It must be that in some cases there is no subalgebra A of Γ which properly contains φ(S), i.e., φ(S) = Γ. In this event, if S contains a δ-idempotent element, then by Theorem 61, S corresponds to a maximal subgroup of $(Z_{2^m d},0)$ and thus $S = \langle q\rangle$ for some prime divisor q of $2^m d$. If q is odd then $\phi(S) \neq \Gamma$, so we must have $S = \langle 2\rangle = E$. On the other hand if S contains no δ- idempotent elements then S does not contain 0, and if in addition S contains no odd elements, then S would be a proper subalgebra of E contrary to its maximality in Ω.

So S must contain at least one odd element b . We must have $b \in Q = \{4t+1 : t \in Z_n, (n = 2^m d)\}$ or $b \in R = \{4t-1 : t \in Z_n, (n = 2^m d)\}$. Suppose $b \in Q$ and $b \notin \langle p\rangle$ for any prime divisor p of d, then gcd(b,n) = 1 and there exists an element c in Q such that bc = 1. Then by Lemma 60, $1 \in bQ = \langle b\rangle_\delta \subseteq S$, with the consequence that $Q = \langle 1\rangle_\delta = D \subseteq S$. The maximality of both S,D in Ω forces S=D.

On the other hand, if $b \in R$, (and $b \notin \langle p\rangle$ for any prime divisor p of d), then $-b \in Q$ and there exists $g \in Q$ such that $-bg = 1$. The mapping $\mu: Q\to bQ$ defined by $\mu(q) = bq$ is a bijection since $\mu(q_1) = \mu(q_2)$ implies $bq_1 = bq_2$, $-g\, bq_1 = -gbq_2$, $q_1 = q_2$, and thus $|bQ| = |Q| = |R|$. Since $b \in R$ and R is a subalgebra of Ω, we have $\langle b\rangle_\delta = bQ \subseteq R$ and thus bQ = R. So S contains R, and since S is maximal in Ω, it must contain some point of Ω not in R, i.e., either S contains a point in E or S contains a point in Q. If S contains a point in Q, then the maximality of R in D allows us to deduce once again that S = D. Suppose S contains a point e in E, then $\delta(-1,e,e) = -2e+1 \in S$. But $-2e+1 \in Q$ and by what was just shown we must have $S \supseteq D$. Actually, since $e \in S$, S properly contains D, and since D is maximal in Ω, we obtain S = Ω, contradicting the maximality of S.

Finally suppose $b \in \langle p\rangle$ for some prime divisor p of d, (and $b \neq 0$), then $\langle b\rangle_\delta = \langle b\rangle = \langle p\rangle$, and thus $0 \in \langle p\rangle \subseteq S$, i.e., S contains a point in E and we obtain the same contradiction as in the previous paragraph. □

Next we do the same for a $(Z_{2^m d},2)$ circle system.

Theorem 71 *Let Ω be a $(Z_{2^m d},2)$ circle system, m ≥2, d odd, and let E, D denote the even and odd vertices of Ω respectively. Then E, D are maximal subalgebras of Ω each equivalent to a $(Z_{2^{m-1} d},1)$ circle system. If p is a prime divisor of d, then Ω contains a maximal subalgebra of the form q+⟨p⟩, (where 4q= 2 (mod 2p)), equivalent to a $(Z_{\frac{2^m d}{p}},2)$ circle system. These constitute the full set of maximal subagebras of Ω.*

Proof That E, D are maximal subalgebras, each equivalent to a $(Z_{2^{m-1} d},1)$ circle system is proved in Theorem 69. We can partition the subalgebras of Ω into three classes: the subalgebras of E, which we call *even,* the subalgebras of D, which we call *odd*, and the subalgebras of Ω which contain both even and odd elements which we call *central.* It is only the maximal subalgebras of Ω in this last class which must be found. If A is a central subalgebra, then A is the disjoint union of the even subalgebra $A\cap E$ and the odd subalgebra $A\cap D$. If $e, f \in A$ with e even and f odd, then the mapping $\mu: A\cap E \to A\cap D$ defined by $\mu(x) = \delta(e,f,x)$ is injective and thus $|A\cap E| \leq |A\cap D|$. Similarly the mapping η: $A\cap D \rightarrow A\cap E$ defined by $\eta(x) = \delta(e,f,x)$ is injective with the consequence that $|A\cap D| \leq |A\cap E|$. We can conclude that every central subalgebra of Ω is the (disjoint) union of an odd and an even subalgebra, each with the same cardinality.

If A is a central subalgebra which is maximal in Ω, then it must be the union of a maximal subalgebra of E and a maximal subalgebra of D. In Theorem 69 it was shown that E and D are each equivalent to $(Z_{n/2}, 1)$ circle systems, and in Theorem 67 the full set of maximal subalgebras of a $(Z_{n/2}, 1)$ circle system is described. To obtain a description of A, it is only necessary to exhibit the correspondence between $A\cap D$, $A\cap E$ and the maximal subalgebras of $(Z_{n/2}, 1)$.

Let Λ denote a $(Z_{n/2}, 1)$ circle system. From Theorem 67, the maximal subalgebras of Λ have the form $s + \langle p\rangle$ where p is an odd prime divisor of n and $4s = 1 \pmod p$. This implies that $p = 4s-1$ or $3p = 4s-1$. Define a mapping ψ: $s + \langle p\rangle \rightarrow E$ by $\psi(s+pt) = 2s+2pt$ and suppose $p = 4s-1$. Then

$\psi(\delta(s+t_1p, s+t_2p, s+t_3p)) = \psi(1 - 3s + p(-t_1-t_2-t_3)) = \psi(1 - 4s + s + p(-t_1-t_2-t_3)) =$

$\psi(s+ p(-t_1-t_2-t_3-1)) = 2s +2p(-t_1-t_2-t_3-1)$, and $\delta(\psi(s+t_1p), \psi(s+t_2p), \psi(s+t_3p)) =$

$\delta(2s+2pt_1, 2s+2pt_2, 2s+2pt_3) = 2 - 6s + 2p(-t_1-t_2-t_3) = 2 - 8s + 2s +2p(-t_1-t_2-t_3) =$

$2s + 2p(-t_1-t_2-t_3-1)$,

so ψ is a homorphism if $p = 4s-1$.

Now suppose $3p = 4s-1$. Then

$\psi(\delta(s+t_1p, s+t_2p, s+t_3p)) = \psi(1 - 4s + s + p(-t_1-t_2-t_3)) = \psi(s + p(-t_1-t_2-t_3 -3)) =$

$2s + 2p(-t_1-t_2-t_3-3)$, and $\delta(\psi(s+t_1p), \psi(s+t_2p), \psi(s+t_3p)) = 2 - 8s + 2s +2p(-t_1-t_2-t_3) =$

$2s + 2p(-t_1-t_2-t_3-3)$.

So in either case ψ is a homorphism. Since also ψ is clearly one to one, we can conclude that ψ is injective.

Now define a mapping ϕ: $s + \langle p\rangle \rightarrow D$ by $\phi(s+pt) = 1 - 2s - 2pt$ and suppose $p = 4s-1$. Then

$\phi(\delta(s+t_1p, s+t_2p, s+t_3p)) = \phi(s+ p(-t_1-t_2-t_3-1)) = 1 - 2s + 2p(t_1+t_2+t_3+1)$,

and $\delta(\phi(s+t_1p), \phi(s+t_2p), \phi(s+t_3p)) = \delta(1\text{-}2s\text{-}2pt_1, 1\text{-}2s\text{-}2pt_2, 1\text{-}2s\text{-}2pt_3) =$

$2 + 6s - 3 + 2p(t_1+t_2+t_3) = 1 + (8s-2) - 2s + 2p(t_1+t_2+t_3) = 1- 2s + 2p(t_1+t_2+t_3+1)$.

If $3p = 4s-1$ then we obtain

$\phi(\delta(s+t_1p, s+t_2p, s+t_3p)) = \phi(s+ p(-t_1-t_2-t_3-3)) = 1 - 2s +2p(t_1+t_2+t_3 +3)$,

and $\delta(\phi(s+t_1p), \phi(s+t_2p), \phi(s+t_3p)) = \delta(1\text{-}2s\text{-}2pt_1, 1\text{-}2s\text{-}2pt_2, 1\text{-}2s\text{-}2pt_3) =$

$2 + 6s - 3 + 2p(t_1+t_2+t_3) = 1 + (8s-2) - 2s + 2p(t_1+t_2+t_3) = 1 - 2s + 2p(t_1+t_2+t_3 +3)$.

Thus ϕ is also an injective homomorphism. We can now conclude that $\psi(s + \langle p\rangle) = 2s + \langle 2p\rangle$ and $\phi(s + \langle p\rangle) = 1 - 2s + \langle 2p\rangle$ are maximal subalgebras of E, D respectively, and that a maximal central subalgebra of Ω must have the form $\{2s + \langle 2p\rangle\} \cup \{1 - 2s + \langle 2p\rangle\}$. If $p = 4s-1$ then $2s = 1-2s+p$ so we obtain $\{2s + \langle 2p\rangle\} \cup \{1 - 2s + \langle 2p\rangle\} = (1-2s) + \langle p\rangle$. If $3p= 4s-1$ then $2s = 1-2s+3p$ and we again obtain $\{2s + \langle 2p\rangle\} \cup \{1 - 2s + \langle 2p\rangle\} = (1-2s) + \langle p\rangle$. Thus every maximal central subalgebra of Ω has the form $(1-2s) + \langle p\rangle$ where p is an odd prime divisor of n and $4s = 1 \pmod p$.

Finally, to show these maximal central subalgebras are $(Z_{\frac{2^m d}{p}}, 2)$ circle systems note that $4(1-2s) = 2 + (2-8s) = 2-6p$ if $3p = 4s-1$, and $4(1-2s) = 2 + (2-8s) = 2 - 2p$ if $p = 4s-1$. Thus if we set $q = 1-2s$ then we always have $4q = 2 \pmod{2p}$. Define the mapping μ: $q+\langle p\rangle \rightarrow (Z_{\frac{2^m d}{p}}, 2)$ by $\mu(q+kp) = -k$. If $(q+pk_1) + (q+pk_2) + (q+pk_3) + (q+pk_4) = 2$ in Ω, then $4q - 2 + p(k_1+ k_2 + k_3+ k_4) = 0 \bmod(2^m d)$ yields $p(k_1+ k_2 + k_3+ k_4+2) = 0 \bmod(2^m d)$, and thus $k_1+ k_2 + k_3+ k_4 = -2$. Then $\mu(q+pk_1) + \mu(q+pk_2) + \mu(q+pk_3) + \mu(q+pk_4) = -k_1- k_2 - k_3- k_4 = 2$ shows that μ yields a $(Z_{\frac{2^m d}{p}}, 2)$ circle system. □

In Figure 51 is illustrated the Hasse diagram formed by the subalgebras of a $(Z_{180},2)$ circle system. Note that the even subalgebras, the odd subalgebras and the central subalgebras form isomorphic lattices.

Figure 51: Subalgebra poset for $(Z_{180}, 2)$ circle system

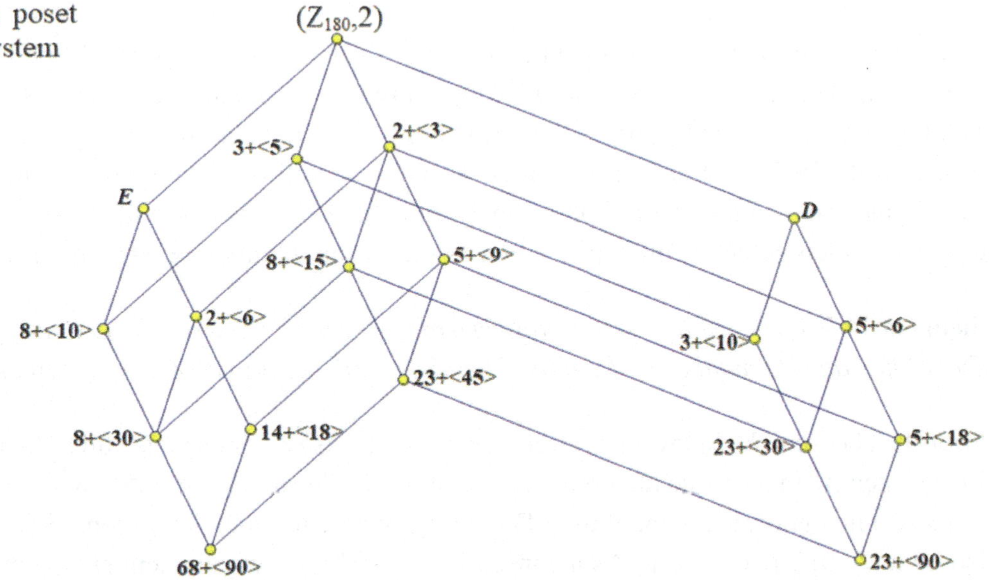

Now having considered all the possibilities for circle systems we have found the following.

Theorem 72 *Every subalgebra of a finite circle system with base Z_n which lies entirely either on the bell or the oval of a self-inversive cubic curve λ of Type IV is again a circle system with base Z_m where $m|n$. The same holds if λ is Type V. All such subalgebras are either subgroups of Z_n or cosets of subgroups of Z_n.* □

In light of Theorem 58 we also obtain

Theorem 73 *Let λ denote either the bell or the oval of a Type IV self-inversive cubic. Then every finite subalgebra S of (λ, δ) is a circle system with base Z_n. The same holds in case λ is a Type V self-inversive cubic.*

Proof Since Theorem 58 involves only finite *cyclic* subalgebras, it remains only to consider those finite subalgebras S such that each component of S_φ is a paracycle. These have the form $S = \varphi^{-m}(T)$ where T is a finite cyclic subalgebra. Suppose T is a subalgebra of the oval of a Type IV cubic λ, then from Theorem 72 we know that T is a circle system with base Z_q for some positive integer q and with sum 0, 1 or 2. We will show that S is a circle system with base $Z_{m^3 q}$ and with the same sum as T.

Let Ω be a ($Z_{m^3 q}$, g) where gcd(q,3) = 1 and where we can suppose that g ∈ {0,1,2}. Let Γ denote a (Z_q,g) circle system. Consider: $\varphi^m : \Omega \to \Omega$ is an endomorphism, and $\varphi^m(\Omega)$ is the cyclic core of Ω. If x, y ∈ Ω then $\varphi^m(x) = t_m g - 3^m x$ if m is odd, and $\varphi^m(x) = t_m g + 3^m x$ if m is even, where t_m is an integer depending on m. Thus $\varphi^m(x) = \varphi^m(y)$ implies either $3^m x = 3^m y \pmod{3^m q}$ or $-3^m x = -3^m y \pmod{3^m q}$. In either case we obtain $x = y \pmod q$. Thus the mapping $\psi: \varphi^m(\Omega) \to \Gamma$ defined by $\psi(x) = x \pmod q$ is a well-defined isomorphism. Now if we let T = Γ, then we must have $S = \varphi^{-m}(T)$ isomorphic to Ω, and thus S is also a circle system with base Z_n, ($n = 3^m q$). A Type V cubic has the same algebraic properties as the bell of a Type IV cubic, so the Theorem also holds in case λ is Type V. □

7.5 Split Subalgebras

It remains to consider *split* subalgebras, i.e., those subalgebras (λ,δ) of a Type IV cubic λ which contain points in the bell as well as the oval. If e, f are any two δ – idempotent elements of such a curve, then the mapping ψ:λ→λ defined by ψ(x) = δ(e,f,x) is easily seen to be an involutory automorphism of (λ,δ). If e is chosen on the bell and f on the oval, then ψ maps the bell onto the oval and the oval onto the bell. Thus the bell and the oval are isomorhphic subalgebras of (λ,δ). First we will consider a split subalgebra which contains an idempotent element of the bell as well as an idempotent element of the oval.

Theorem 74 *Let S be a finite split subalgebra of (λ,δ) where λ is a self-inversive Type IV cubic, and suppose S∩oval has an odd number n of points. Then S is equivalent to a ($Z_2 \times Z_n$,(0,0)) circle system.*

Proof By Theorem 73, S∩oval is a circle system with base Z_n. Since n is odd S∩oval is equivalent to a (Z_n,0) circle system with a unique idempotent element f = 0. The mapping ψ: S∩oval → S∩bell defined by ψ(x) = δ(e,f,x) is an isomorphism and thus ψ(f) = e must be the unique idempotent element of S∩bell which must also be a (Z_n,0) circle system. As a consequence {e,f} is a two element subalgebra of S. Let Γ denote the (Z_n,0) circle system on the oval, and let Λ denote the (Z_n,0) circle system on the bell.

If X, Y are the centers of the two inverters on the bell, then every idempotent element of λ squares to either X or Y. Suppose f*f = Y, then e = δ(e,f,f) = Y*{(Y*e)*(f*f)} = Y*{(Y*e)*Y} = Y*e, and thus e*e = Y. Now relabel the points of Γ by k →(1,k) and choose for e the label (0,0). (e can be either one of the two idempotent elements on the bell which square to Y.) The remaining points of Λ are determined by ψ(1,k) = δ((0,0),(1,0), (1,k)) to which we give the label (0,-k). We will show that the points of S so labeled form a ($Z_2 \times Z_n$,(0,0)) circle system Ω. It suffices by Theorem 34 to show that the points of Ω constitute the vertex set of a perfect $Z_2 \times Z_n$ – gon. Also by Theorem 34, both Γ, Λ are perfect polygons. In particular, if (1,s), (1,t) belong to Γ, then (1,s)*(1,t) = $X_{(0,s+t)}$. Consider:

$$(0,s)*(0,t) = \delta((0,0),(1,0),(1,-s)) * \delta((0,0),(1,0),(1,-t)) =$$

$$\{Y*\{(Y*(0,0))*((1,0)*(1,-s))\}\} * \{Y*\{(Y*(0,0))*((1,0)*(1,-t))\}\} =$$

$$\{Y*\{(0,0)*((1,0)*(1,-s))\}\}* \{Y*\{(0,0)*((1,0)*(1,-t))\}\} =$$

$$(Y*Y)* [\{(0,0)*((1,0)*(1,-s))\}* \{(0,0)*((1,0)*(1,-t))\}] =$$

$$(Y*Y)*[\{(0,0)*(0,0)\}*\{((1,0)*(1,0))* ((1,-s)* (1,-t))\}] =$$

$$(Y*Y)*\{Y*\{Y*((1,-s)*(1,-t))\}\} = (Y*Y)* ((1,-s)*(1,-t)) = (Y*(1,-s))* (Y*(1,-t)) =$$

$$(X_{(0,0)}*(1,-s))* (X_{(0,0)}*(1,-t)) = (1,s)*(1,t) = X_{(0,s+t)}.$$

Thus the perspective points of Γ are the same as the perspective points of Λ. But to show that Ω is a perfect $Z_2 \times Z_n$ – gon, it remains to show the existence of perspective points of the form $X_{(1,s)}$. Define $X_{(1,s)} = (0,0)*(1,s)$ for each $s \in Z_n$, and suppose $p+q = s$ with $p,q \in Z_n$. Consider:

$$(1,q)*(0,p) = \{(1,-s)*X_{(0,q-s)}\}*\delta((0,0),(1,0),(1,-p)) =$$

$$\{(1,-s)*X_{(0,q-s)}\}*\{Y*\{(Y*(0,0))*((1,0)* (1,-p))\} = \{(1,-s)*X_{(0,q-s)}\}*\{Y*\{(0,0)* ((1,0)*(1,-p))\} =$$

$$(Y*(1,-s))* \{X_{(0,q-s)}*\{(0,0)* ((1,0)*(1,-p))\} = (1,s)*\{X_{(0,q-s)}*\{(0,0)*X_{(0,-p)}\}\} =$$

$$(1,s)*\{X_{(0,q-s)}* (0,-p)\} = (1,s)*(0, q+p-s) = (1,s)*(0,0) = X_{(1,s)}.$$

Thus all lines [(1,q),(0,p)] with $p+q = s$ concur at $X_{(1,s)}$, showing that $X_{(1,s)}$ is indeed a perspective point of Ω. Then Ω is a perfect $Z_2 \times Z_n$ – gon with all vertices and perspective points on the Type IV self-inversive cubic λ. Also since $Y = X_{(0,0)}$ is the center of a circle which inverts λ onto itself, it follows from Theorem 34 that Ω, and thus S, is a $(Z_2 \times Z_n,(0,0))$ circle system. □

The following two Lemmas will be useful for considering the case in which the split subalgebra contains an even number of points in the oval.

Lemma 75 *Let S be a finite split subalgebra of (λ,δ) where λ is a self-inversive Type IV cubic. Then S∩oval and S∩bell have the same number of points.*

Proof Let a ∈ S∩oval and b ∈ S∩bell. Then the mapping μ: S∩oval → S∩bell defined by μ(x) = δ(x,a,b) is easily seen to be a bijection. □

Lemma 76 *Let points E, F, A lie on the oval of a Type IV cubic λ, and let B = δ(E,A,F). Let point G lie on the bell of λ. Then δ(G,A,B) = G iff E*F = G*G.*

Proof Suppose E*F = G*G, then δ(G,A,B) = δ(G,A,δ(E,A,F)). Consider: δ(E,A,F)) = Y*{(Y*A)*(E*F)} = Y*{(Y*A)*(G*G)} = δ(A,G,G) and thus δ(G,A,B) = δ(G,A, δ(A,G,G)) = G. Conversely, suppose δ(G,A,B) = G, then Y*{(Y*A)*(E*F)} = B = δ(A,G,G) = Y*{(Y*A)*(G*G)}, and E*F = G*G follows by cancellation. □

Theorem 77 *Let S be a finite split subalgebra of (λ,δ) where λ is a self-inversive Type IV cubic, and suppose S∩oval is a (Z_n,g) circle system Γ with n even. Then S∩bell is also a (Z_n, g) circle system and S is equivalent to either a ($Z_2 \times Z_n$,(0,g)) or a (Z_{2n}, 2g) circle system.*

Proof Let Λ = S∩bell, let x ∈Λ and let q be a fixed element of Γ. By Lemma 75, Λ, Γ both have n points and thus the one to one mapping ψ: Γ→Λ defined by ψ(s) = δ(x,q, g-s-q) is onto, and thus Λ = {δ(x, q, g-s-q) : s ∈Γ}. Since x∈Λ there exists s∈Γ such that δ(x,q, g-s-q) = x.

Suppose s is even and relabel the points of Γ by k → (1,k). Then we have δ(x,(1,q), (1, g-s-q)) = x, and we give x the label x = (0,s/2). Now apply Lemma 76 with G = x; E = (1,0); F = (1,s); A = (1,q) and B = δ(E,A,F) = δ((1,0),(1,q),(1,s)) = (1, g-s-q) to obtain x*x = (1,0)*(1,s) = $X_{(0,s)}$. Label the remaining points of Λ by (0,t) = δ(x, (1,0), (1, g-t- $\frac{S}{2}$).

First we show that Λ,Γ have the same perspective points. Suppose (0,p), (0,t) ∈Λ and (1,a), (1,b) ∈Γ with p+t = a+b. We must show that (0,p)*(0,t) = (1,a)*(1,b). Consider:

$$(0,p)*(0,t) = \delta(x,\ (1,0),\ (1,\ g\text{-}p\text{-}\ \tfrac{S}{2}))\ *\ \delta(x,\ (1,0),\ (1,\ g\text{-}t\text{-}\ \tfrac{S}{2})) =$$

$$[Y*\{(Y*x)*((1,0)*(1,\ g\text{-}p\text{-}\ \tfrac{S}{2}))\}]\ *\ [Y*\{(Y*x)*((1,0)*(1,\ g\text{-}t\text{-}\ \tfrac{S}{2}))\}] =$$

$$(Y*Y)\ *\ [\{(Y*Y)*(x*x)\}\ *\ \{((1,0)*(1,0))\ *\ ((1,\ g\text{-}p\text{-}\ \tfrac{S}{2})\ *\ (1,\ g\text{-}t\text{-}\ \tfrac{S}{2}))\}].$$

From δ(x,(1,q), (1, g-s-q)) = x we obtain Y*{(Y*x)*$X_{(0,g\text{-}s)}$} = x which yields (Y*Y)*(x*x) = $X_{(0,g\text{-}s)}$, so we have

$$(0,p)*(0,t) = (Y*Y)\ *\ [X_{(0,g\text{-}s)}\ *\ \{((1,0)*(1,0))\ *\ X_{(0,\ 2g-p-t-s)}\}] =$$

$$(Y*Y)\ *\ [((1,g\text{-}s)*(1,0))\ *\ \{((1,0)*(1,0))\ *\ ((1,g\text{-}s)*(1,g\text{-}a\text{-}b))\}] =$$

$$(Y*Y)\ *\ [(1,0)*(1,\ g\text{-}a\text{-}b)].$$

Now from (Y*Y)*(x*x) = $X_{(0,g\text{-}s)}$ and x*x = (1,0)*(1,s) we obtain Y*Y = ((1,0)*(1,s))* $X_{(0,g\text{-}s)}$ so we have

$$(0,p)*(0,t) = [((1,0)*(1,s))*X_{(0,g\text{-}s)}]\ *\ [(1,0)*(1,\ g\text{-}a\text{-}b)] = (1,s)\ *\ (1,\ a+b-s) =$$

$$X_{(0,\ a+b)} = (1,a)*(1,b).$$

To show that the points of S form the vertex set of a $Z_2 \times Z_n$–gon, it remains to prove perspective points of the form $X_{(1,s)}$. Let (0,p), (0.t) ∈Λ and (1,r), (1,u) ∈Γ with p+r = t+u. Consider:

$$(0,p)*(1,r) = (1,r)\ *\ \delta(x,\ (1,0),\ (1,\ g\text{-}p\text{-}\ \tfrac{S}{2})) =$$

$$(1,r)*[Y*\{(Y*x)*X_{(0,\ g-p-s/2)}\}] = [Y*(Y*(1,r)]\ *\ [Y*\{(Y*x)*X_{(0,\ g-p-s/2)}\}] =$$

$$(Y*Y)\ *\ [(Y*(1,r))*\{(Y*x)*X_{(0,\ g-p-s/2)}\}] = (Y*Y)\ *\ [x\ *\ ((1,r)*X_{(0,\ g-p-s/2)})] =$$

$$(Y*Y)\ *\ [x\ *\ (1,\ g\text{-}p\text{-}r\text{-}\ s/2)] = (Y*Y)\ *\ [x\ *\ (1,\ g\text{-}t\text{-}u\text{-}\ s/2)].$$

The same estimate is obtained for (0,t)*(1,u) by just replacing p by t and r by u in the above sequence of equations. We thus obtain that S is a perfect $Z_2 \times Z_n$ –gon, and by Theorem 34, S is a $(Z_2 \times Z_n,(0,g))$ circle system.

Now suppose s is odd. We have δ(x, q, g-s-q) = x; relabel the points of Γ by k→2k so we have δ(x, 2q, 2g-2s-2q) = x. Now we give x the label x = s, and applying Lemma 76 with G = x; A = 2q; E = 0; F = 2s and B = 2g-2s-2q we have δ(G,A,B) = δ(x, 2q, 2g-2s-2q) = x = G, and thus x*x = 0*2s = X_{2s}. The remaining, (odd), points of Λ are labeled using the formula q = δ(x,0, 2g-x-q). Following the familiar program we first show that Γ,Λ have the same perspective points. Let p, u ∈Λ and let 2r, 2w ∈Γ with p+u = 2r+2w. Consider:

$$p*u = \delta(x,0,\ 2g\text{-}x\text{-}p) * \delta(x,0,\ 2g\text{-}x\text{-}u) =$$

$$[Y*\{(Y*x)*(0*(2g\text{-}x\text{-}p))\}] * [Y*\{(Y*x)*(0*(2g\text{-}x\text{-}u))\}] =$$

$$(Y*Y) * [((Y*Y)*(x*x)) * ((0*0) * X_{(4g\text{-}2x\text{-}p\text{-}u)})].$$

Now from $\delta(x,\ 2q,\ 2g\text{-}2s\text{-}2q) = x$ *we obtain* $(Y*Y)*(x*x) = X_{(2g\text{-}2s)}$, *and thus*

$$p*u = (Y*Y) * [X_{(2g\text{-}2s)} * \{(0*0)* X_{(4g\text{-}2x\text{-}p\text{-}u)}\}] =$$

$$(Y*Y) * [(0*(2g\text{-}2s)) * \{(0*0)* X_{(4g\text{-}2s\text{-}2r\text{-}2w)}\}] = (Y*Y) * [0 * (2g\text{-}2r\text{-}2w)] =$$

$$\{(x*x)* X_{(2g\text{-}2s)}\} *\{0 * (2g\text{-}2r\text{-}2w)\} = \{(0*2s) * X_{(2g\text{-}2s)}\} *\{0 * (2g\text{-}2r\text{-}2w)\} =$$

$$2s * (\text{-}2s + 2r + 2w) = X_{2r+2w} = 2r*2w.$$

This proves that Γ,Λ have the same perspective points, all of which have even subscripts. To show that S is a perfect Z_{2n} – gon it remains to prove the perspective points with odd subscript. Let r, u ∈Λ and 2p, 2q ∈ Γ with r+2p = u+2q. We must show that r*2p = s*2q. Consider:

$$r*2p = \delta(x,\ 0,\ 2g - x\ \text{-}r) * 2p = [Y * \{(Y*x)* X_{2g-x--r}\}] * 2p =$$

$$[Y * \{(Y*x)* X_{2g-x-r}\}] * (Y * (Y*2p)) = (Y*Y) * [(Y*2p) * \{(Y*x)* X_{2g-x-r}\}] =$$

$$(Y*Y) * [x * (2p * X_{2g-x-r}\}] = (Y*Y) * [x * (2g - x - r - 2p)] = (Y*Y) * [x * (2g - x - u - 2q)].$$

Now putting u in place of r and q in place of p in the above sequence, the same result is obtained. The points of S form a perfect Z_{2n} – gon, and from Theorem 34 we obtain that S is a $(Z_{2n},\ 2g)$ circle system. □

The construction of Theorem 77 is illustrated in Figure 52 with n = 4 and s = q = 0 and in Figure 53 with n = 4 and s = q = 1. Combining the results of Theorems 6, 73, 74, 77 we obtain the following.

Theorem 78 *Let λ be an irreducible self-inversive cubic curve. Then every finite subalgebra of (λ,δ) is equivalent to a* $(Z_n,0)$, $(Z_n,1)$, $(Z_n,2)$, $(Z_2 \times Z_n,(0,0))$, $(Z_2 \times Z_n,(0,1))$, *or a* $(Z_2 \times Z_n,(0,2))$ *circle system.* □

Note that with Theorem 78 we have obtained a second proof that $(Z_2 \times Z_n,(1,0))$ and $(Z_2 \times Z_n,(1,2))$ circle systems, (see Sections 6.3, 6.4), do not exist with a self-inversive cubic envelope.

Figure 52: Split (Z_2 x Z_4,(0,1)) circle system obtained as extension of (Z_4 ,1) circle system on oval

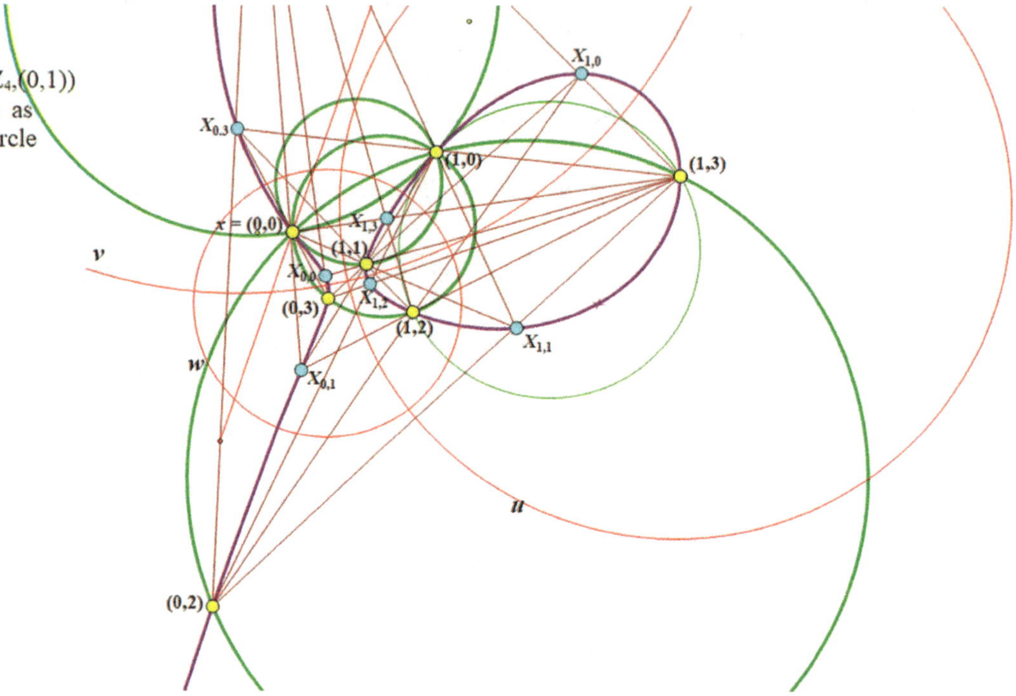

Figure 53: (Z_8 ,2) circle system obtained as extension of (Z_4 ,1) circle system on oval

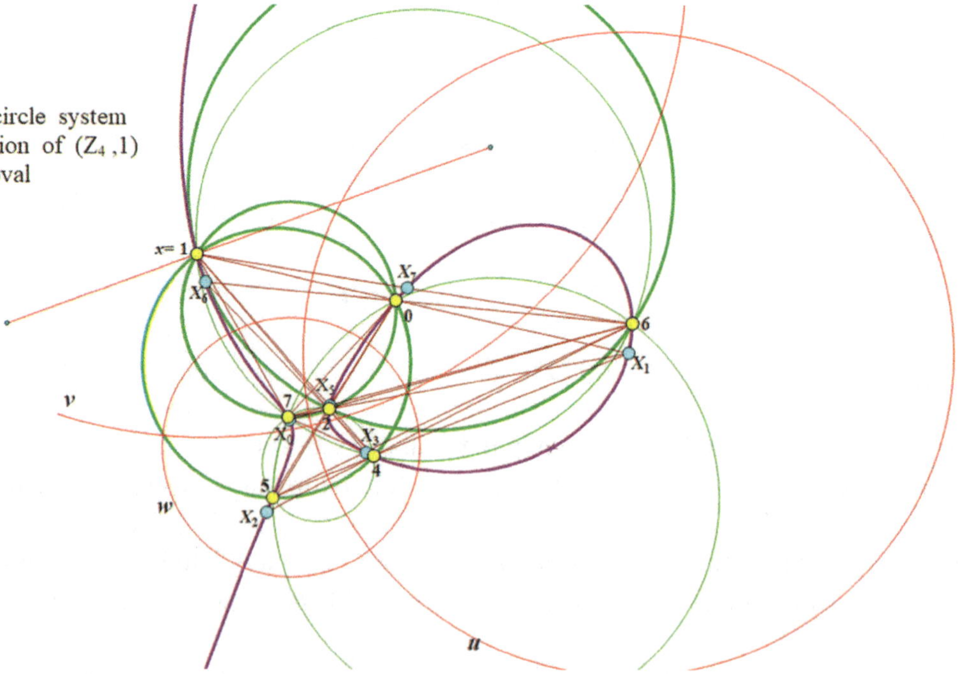

Chapter 8

Circle Systems on Conics

8.1 A Parallelism property for Noncircular conics

We have thus far considered only circle systems whose envelope is a self-inversive cubic or the inverse of such a curve. In [4] it is shown that circle systems can also be constructed on conics. For the sake of completeness we will repeat those results in this chapter. If α is a noncircular conic and we invert α by a circle with center on α, then a self-inverseive cubic curve α' is obtained. In case α is a parabola, then the axis of symmetry of α is transformed to a self-inverting circle for α', and in case α is an ellipse or hyperbola, then the two perpendicular axes of symmetry are transformed into orthogonal self-inverting circles for α'. In the case of an ellipse α' is a Type V cubic so we can construct circle systems with base Z or Z_n on α' using the methods described in earlier chapters. By inverting back to α, we obtain such circle systems on an ellipse. In case α is a parabola or a hyperbola, then α' is singular and it becomes impossible to construct finite justified circle systems on α', although circle systems with base Z can be constructed on α' using the construction given below. Our constructions in this chapter will be directly on the noncircular conic α without any recourse to the cubic α'. First we have a basic result concerning the intersection of a circle with a parabola. This will be useful when we discuss the intersection of a circle with a general noncircular conic. Basic properties of conics can be found in [3].

Lemma 79 *Let p denote the parabola (1) $y = a(x-h)^2 + k$, and let c be a circle such that $p \cap c$ contains the four points $A=(s,t)$, $B = (q,p)$, $C = (u,v)$ and $D = (m,n)$. Then $4h = s+q+u+m$. Moreover, if $A' = (-s,t)$ and $C' = (-u,v)$, then BD is parallel to A'C'.*

Proof Let (2) $(x-d)^2 + (y-e)^2 = r^2$ be the equation of c and use (1) to substitute for y in (2) to obtain (3) $(x-d)^2 + (a(x-h)^2 + k - e)^2 = r^2$. The coefficient of x^3 in (3) is $-4a^2h$. Since the points A,B,C,D satisfy (3), (3) must have the form: (4) $a^2(x-s)(x-q)(x-u)(x-m)$. The coefficient of x^3 in (4) is $a^2(-s-q-u-m)$, so we must have $-4a^2h = a^2(-s-q-u-m)$ which yields $4h = s+q+u+m$.

To prove the last statement of the Lemma we simply compute the slopes:

$$M_{BD} = \frac{p-n}{q-m} = \frac{(a(q-h)^2+k)-(a(m-h)^2+k)}{q-m} = a(q+m-2h),$$

$$M_{A'C'} = \frac{t-v}{-s+u} = \frac{((a(s-h)^2+k)-(a(u-h)^2+k))}{(-s+u)} = a(-s-u+2h).$$

Now by the first part of the Lemma: $M_{BD} = a(q+m-2h) = a(4h-s-u-2h) = M_{A'C'}$. □

The parallelism property stipulated in Lemma 79 holds for any noncircular conic as we show next.

Theorem 80 *Let A,B,C,D be four points on a noncircular conic α with A,B,C distinct, and let A', C' be the reflections of A, C respectively across an axis of symmetry of α. Then A,B,C,D are cocyclic iff BD is parallel to A'C'.*

Proof Suppose A,B,C,D lie on a circle c with equation: (5) $(x-h)^2 + (y-k)^2 = r^2$. If α is a parabola, then BD is parallel to A'C' by Lemma 79, so we may suppose that α is an ellipse with equation: (6) $x^2/a^2 + y^2/b^2 = 1$, or a hyperbola with equation: (7) $x^2/a^2 - y^2/b^2 = 1$. Suppose A', C' are reflections of A, C across the vertical axis of symmetry (y-axis) of α. If we solve (6) or (7) along with the equation of c by solving (5) for y^2 and substituting for y^2 in (6) or (7), we obtain the equation of a parabola (with vertical axis of symmetry) which contains the four points A,B,C,D. We may then invoke Lemma 79 to conclude that BD is parallel to A'C'. If A", B" are reflections of A,C across the horizontal axis of symmetry of α, then we obtain similarly that A"C" is parallel to BD. Thus the lines BD, A'C', A"C" are mutually parallel. Conversely, suppose BD is parallel to A'C' and let E be the fourth point on α and on circle (A,B,C). By the first part of the Theorem, A'C' is parallel to BE and thus BE and BD are parallel. We must then have that BE and BD are the same line. Any line through B meets the conic α in at most one point (besides B), so we must have E = D, and thus the points A,B,C,D are cocyclic. □

8.2 Construction of a (Z,0) Circle System on a Noncircular Conic

In this Section we present the construction of a (Z,0) circle system on a noncircular conic. If a, b are elements of the set Z of integers, we denote, (as usual), the line joining the corresponding points by [a,b].

Theorem 81 *Let α be a noncircular conic with axis of symmetry j. In the case of a hyperbola the transverse axis is selected for j so that in all cases j and α have a point in common. The following assignment of integer labels to points on α yields a (Z,0) circle system: (i) Assign the integer 0 to a common point of j and α. (ii) Put vertex 1 on α but not on an axis of symmetry of α. (iii) Let vertex -1 be the reflection of vertex 1 across j, (iv) Let m denote the line [0,-1] and draw a line through vertex 1 and parallel to m. Label the point where this line meets α with -2. (v) Let vertex 2 be the reflection of vertex -2 across j. (vi) In general, if elements {0, ±1, ±2, ..., ±t} have been assigned to points on α, assign -(t+1) to the point on α and on the line through t and parallel to m, and let t+1 be the reflection of -(t+1) across j. We assume that the generating vertex 1 is chosen on α so that no point on α is assigned more than one integer label.*

The construction described in Theorem 81 is illustrated in Figure 54. Before proving Theorem 81 we need the following Lemma. We employ a variant of Pascal's Theorem which states that if hexagon (a,b,c,d,e,f) is inscribed in a conic and two pairs of opposite sides are parallel, then the third pair of opposite sides are also parallel. Pascal's Theorem for a hexagon inscribed in a circle can be found in [8], and the more general case in which the hexagon is inscribed in any conic can be found in [3].

Lemma 82 *In the proposed construction of a (Z,0) circle system on a noncircular conic, if m,n,p,q are integers and m+n = p+q, then the lines [m,n] and [p,q] are parallel.*

Proof Consider the inscribed hexagon (p, q, -q-1, q+1, p-1, -p). Opposite sides [p,-p], [q+1, -q-1] are vertical hence parallel, and opposite sides [q, -q-1], [p-1, -p] are parallel by virtue of our construction. By Pascal's Theorem, the remaining pair of opposite sides [p,q], [p-1, q+1] are also parallel. Applying the same result with p-1 in place of p and q+1 in place of q, we obtain that the lines [p-1, q+1], [p-2, q+2] are parallel, and thus [p,q], [p-2, q+2] are parallel. Repeating this argument we obtain that all lines of the form {[p-t, q+t] : t ∈ Z} are mutually parallel. If we let t = p-m then we obtain that the lines [p,q], [m,n] are parallel. Note that the same proof works in case p = q, when line [p,q] is interpreted as the tangent to the conic at p. □

Figure 54: (Z,0) Circle System on Ellipse

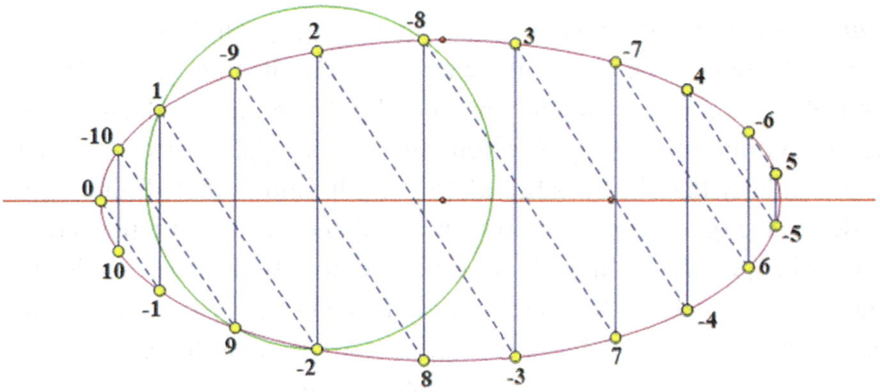

Now we may proceed with the proof of Theorem 81.

Proof (of Theorem 81): Let a,b,c,d be four distinct integers such that a+b+c+d = 0. We must show that the corresponding points on α are cocyclic. Consider: ∠(a,b,c) = ∠(-a,-b,-c) since -a, -b, -c are reflections of a, b, c across the axis of symmetry j. Since (-a) + (-b) = c + d, the lines [-a, -b], [c,d] are parallel by Lemma 82. Also, since (-b)+(-c) = c + (-b – 2c), the lines [-b, -c], [c, -b-2c] are parallel by Lemma 82. It then follows that ∠(-a,-b,-c) = ∠(d, c, -b-2c) as in Figure 54. Since a+d = c + (-b-2c), the lines [a,d], [c, -b-2c] are parallel by Lemma 82. So if we extend segment [c,d] to point e as in Figure 55, we have ∠(a,d,e) = ∠(d,c, -b-2c) = ∠(a,b,c). Consequently ∠(a,d,c) = 180° - ∠(a,d,e) = 180° - ∠(a,b,c), and it follows that quadrilateral (a,b,c,d) is cyclic. (The elementary result that a quadrilateral is cyclic iff its opposite angles are supplementary can be found in [10].) □

Let G be any abelian group and let 0 denote the identity in G. Next we show that any (G,0) circle system on a noncircular conic must exhibit the properties inherent in our construction of a (Z,0) circle system.

Figure 55: Cyclic Quadrilateral in (Z,0) Circle System

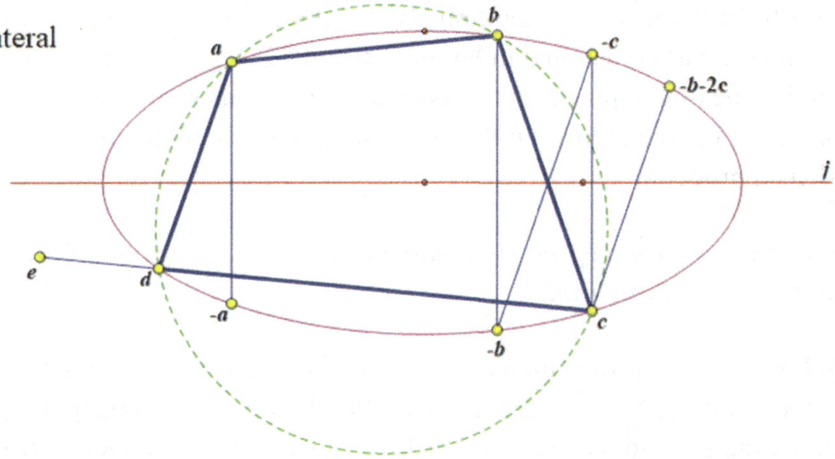

Theorem 83 *Let Ω be a (G,0) circle system whose vertices lie on a noncircular conic α and such that G contains distinct points {g, -g, h, -h, k, -k} . Then one of the axes of symmetry of α has the property that g and -g are reflections with respect to it, for every g $\in$ G. Moreover, if a,b,c,d $\in$ G and a+b = c+d, then [a,b] is parallel to [c,d]. Conversely, if a set Ω of points on a noncircular conic is labeled with the elements of an abelian group G such that whenever a,b,c,d $\in$ G with a+b = c+d, the lines [a,b], [c,d] are parallel, then Ω is a (G,0) circle system on the conic.*

Proof Let g,h,k be elements of G such that {g, -g, h, -h, k, -k} are distinct. Then c_1 = (g, -g, h, -h), c_2 = (h, -h, k, -k) and c_3 = (k, -k, g, -g) are circles in Ω. The lines [g, -g], [h, -h], [k, -k] either meet at the radical center of c_1, c_2, c_3 or are mutually parallel. Let h', -h' denote the reflections of h, -h respectively across an axis j of symmetry of α. By Theorem 80, the lines [g, -g], [h', -h'] are parallel. Considering circle c_2 we obtain similarly that the lines [k, -k], [h', -h']. It follows that all lines {[g, -g] : g $\in$ G, g $\neq$ -g} are mutually parallel. Since [g, -g] is parallel to both [h, -h] and [h', -h'], it follows that [h, -h] is parallel to [h', -h']. This can only happen if h' = -h or [h, -h] is parallel to j. In the first case h and -h are reflections with respect to j, and in the second case h, -h are reflections with respect to a second axis of symmetry of α perpendicular to j. The circles c_4 = (0, g, h, -(g+h)) and c_5 = (0, -g, -h, g+h) belong to Ω and are reflections across an axis of symmetry of α. Since both circles contain vertex 0, we must have that 0 lies on this axis. Now suppose g, -g are reflections across the axis j of symmetry of α for all g $\in$ G, and suppose a,b,c,d $\in$ G such that a+b = c+d. Then a+b+(-c)+(-d) = 0 and so (a,b,-c,-d) is a circle in Ω provided a,b,-c,-d are distinct elements of G. In this case we obtain [a,b] parallel to [c,d] by Theorem 80. If two of these points are identical, say a = b, then we interpret (a,a,-c,-d) as the circle (a, -c, -d) which is tangent to α at a. In this case Theorem 80 says that [a,a] = [a,b] is parallel to [c,d] where [a,a] denotes the tangent to α at a. Conversely, suppose Ω is a set of points on a noncircular conic α, labeled with the elements of an abelian group G such that whenever a,b,c,d $\in$ G and a+b = c+d, the lines [a,b], [c,d] are parallel. Let p,q,r,s $\in$ G and suppose p+q+r+s = 0. Then p+q = (-r) + (-s) and so the lines [p,q], [-r,-s] are parallel. But then, by Theorem 80, the points p,q,r,s are cocyclic and thus Ω is a (G,0) circle system. □

8.3 A Ternary Hypercommutative Algebra on a Noncircular Conic

Let α be a noncircular conic and let a,b,c be three points on α. Circle (a,b,c) must meet α in a fourth point which we denote by δ(a,b,c). This ternary operation is still defined in case two or even three of its arguments are identical; δ(a,a,b) is the remaining point on α and on the circle which is tangent to α at a and contains b, and δ(a,a,a) is the remaining point on α and on the unique circle which intersects α with multiplicity 3 at a. Regarding intersection multiplicity for algebraic curves, the reader may wish to consult [3]. These possibilities are are illustrated in Figures 56 and 57. A point x of α is *δ-idempotent* if δ(x,x,x) = x. It can be shown that the δ-idempotent elements of an ellipse consist precisely of the four points where the major and minor axes meet the ellipse. In this Section we will show that (α,δ) is a THA. We begin with yet another parallelism property.

Lemma 84 *Let a,b,c,d,e,f be six points on a noncircular conic α. Then the lines [δ(a,b,c), δ(d,e,f)] and [δ(a,b,d), δ(c,e,f)] are parallel.*

Proof Let j be an axis of symmetry of α and if x is any point on α, let x' denote the reflection of x across j. Since the points {a,b,c, δ(a,b,c)} are cocyclic, the lines [c, δ(a,b,c)] and [a',b'] are parallel by Theorem 80. Similarly, the lines [d, δ(a,b,d)] and [a',b'] are parallel. Consequently the lines [c, δ(a,b,c)] and [d, δ(a,b,d)] are parallel as are the lines [d, δ(d,e,f)] and [c, δ(c,e,f)]. See Figure 58. If we now consider the hexagon (δ(a,b,c), δ(d,e,f), d, δ(a,b,d), δ(c,e,f), c) inscribed on α, and apply Pascal's Theorem (variant), we obtain the desired result. □

Now we are prepared to show that the ternary hypercommutativity holds on any noncircular conic.

Theorem 85 *Let α be a noncircular conic and let a,b,c,d,e,f,q be any seven points on α. Then δ(δ(a,b,c), δ(d,e,f), q) = δ(δ(a,b,d), δ(c,e,f), q).*

Proof: Let x = δ(δ(a,b,c), δ(d,e,f), q) as in Figure 59, and let j be an axis of symmetry of α. For each point w on α, let w' denote the reflection of w across j. By Theorem 80 we have (8) [x,q] parallel to [δ(a,b,c)', δ(d,e,f)']. Let y = δ((a,b,d), δ(c,e,f), q), then by Theorem 80 we have (9) [y,q] parallel to [δ(a,b,d)', δ(c,e,f)']. By Lemma 84 we have [δ(a,b,c), δ(d,e,f)] parallel to [δ(a,b,c), δ(d,e,f)], and by reflecting these lines across j we obtain (10) [δ(a,b,c)', δ(d,e,f)']

Figure 56: δ-operator with two Identical Arguments

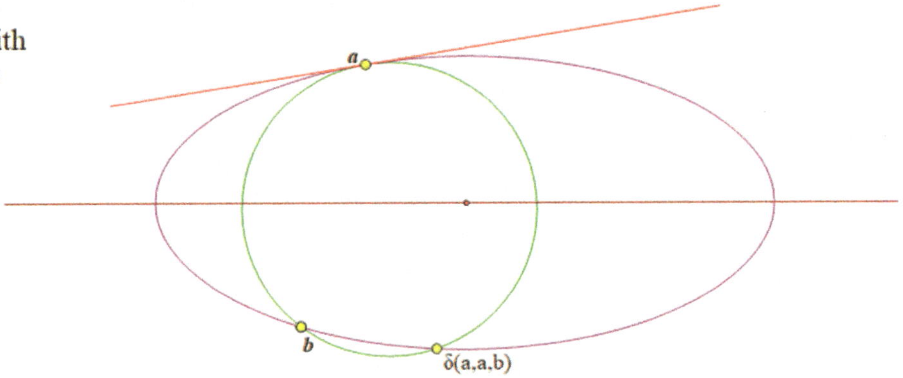

Figure 57: δ-operator with three identical arguments

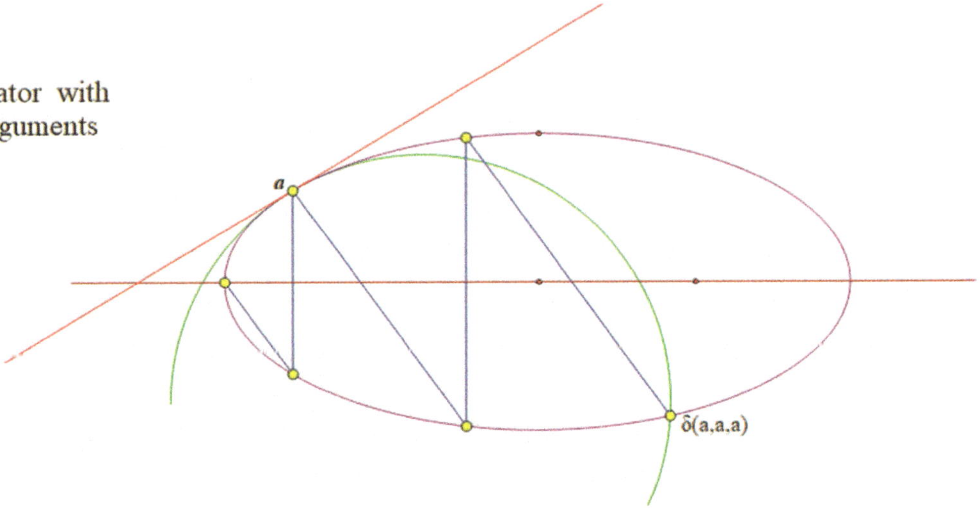

parallel to [δ(a,b,d)', δ(c,e,f)']. From (8),(9), (10) we obtain that the lines [x,q], [y,q] are parallel. This implies that the points x,y,q are collinear. But any line through q meets the conic α in at most one other point, so we must have x = y. □

8.4 Construction of Circle systems with Base Z_n

In the construction of a (Z,0) circle system given in Theorem 81 suppose that the noncircular conic α is an ellipse with x and y axes as axes of symmetry. Let vertex 0 be chosen on the negative x-axis and let $k \neq 2$ be an integer. The placement of vertex 1 on the ellipse determines the positions of the remaining vertices and it is clear, from the construction given in Theorem 81, that vertex 1 can be positioned sufficiently close to vertex 0 so that vertices 1, …, k lie in the upper half of the ellipse and their reflections -1, …, -k lie on the lower half of the ellipse. By shifting vertex 1 a little to the right along the ellipse we can make vertices k and -k coincide at the other extremity of the major axis. This is

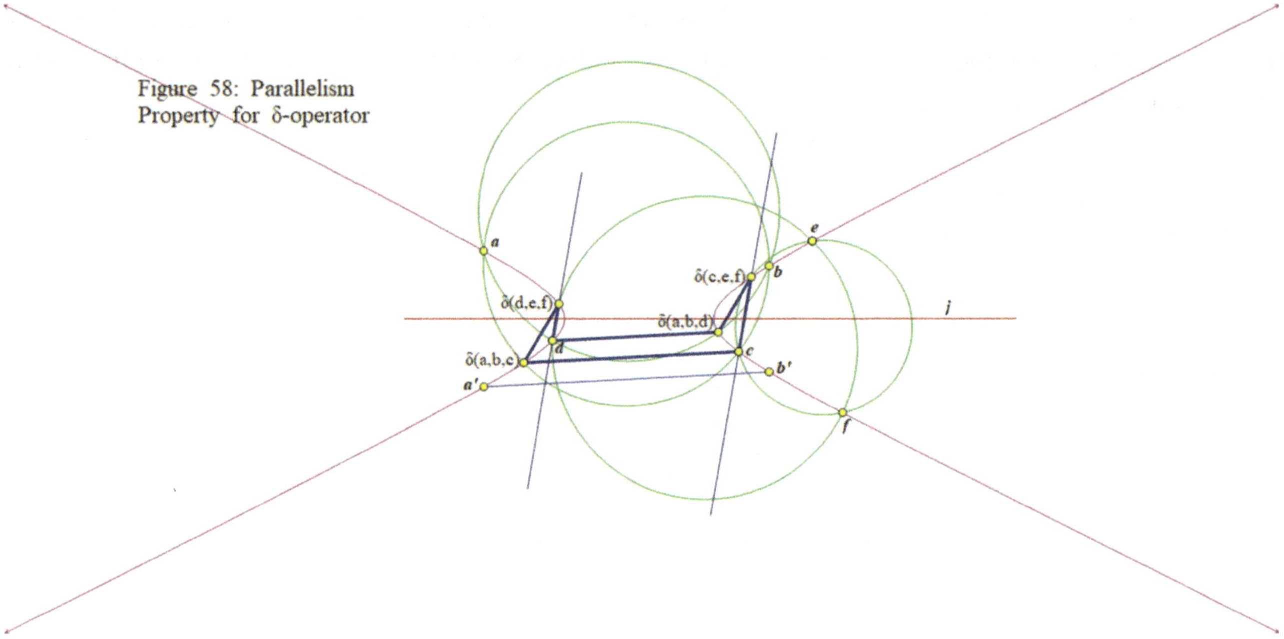

Figure 58: Parallelism Property for δ-operator

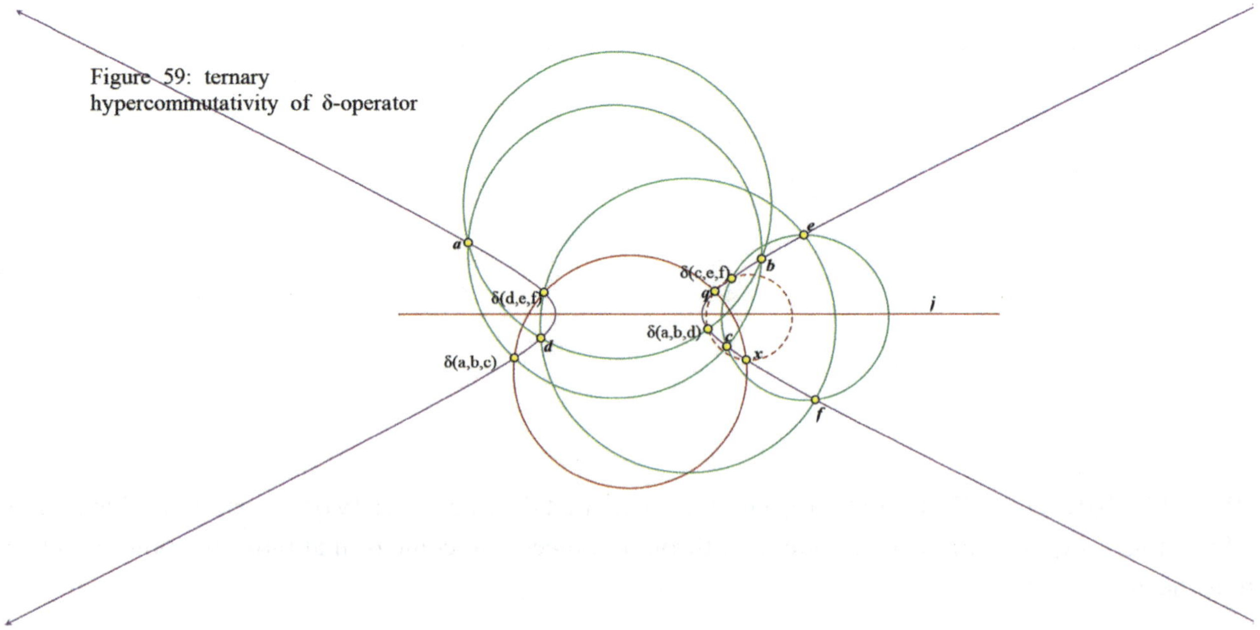

Figure 59: ternary hypercommutativity of δ-operator

illustrated for k = 5 in Figure 60. When this happens each pair of vertices which are congruent mod(2k) also coincide. The result is a $(Z_{2k}, 0)$ circle system on the ellipse. If instead we shift vertex 1 a little to the left we can make vertex k and vertex -(k+1) coincide, resulting in a $(Z_{2k+1}, 0)$ circle system. Note that this construction is not possible on a parabola or hyperbola since these are not closed curves and our construction puts new vertices further and further away from vertex 0. We thus have the following result.

Theorem 86 *For any positive integer n, a $(Z_n, 0)$ circle system can be constructed on any ellipse but not on a parabola or hyperbola.* □

In the opening paragaph of Chapter 4, it was observed that $(Z_{n,} 1)$ and $(Z_{n,} 2)$ circle systems exist as subsystems of a $(Z_{n,} 0)$ circle system. Then, along with Theorem 6 we have

Theorem 87 *Every circle system with base Z_n can be constructed on any ellipse.* □

With regard to the other major class of finite circle systems, i.e., those with base $Z_2 \times Z_{2m}$ we can show that no such circle system can exist on a noncircular conic. If Ω is any (G,g) circle system, recall that a circle of the form (a,b,x,x) where a+b+2x = g is called auxiliary. If the points of Ω lie on a noncircular conic α, then this auxiliary circle is tangent to α at x. Let Ω be a $Z_2 \times Z_{2m}$ circle system, m ≥ 2, and suppose Ω lies on a noncircular conic α. Since $Z_2 \times Z_{2m}$ contains a subgroup isomorphic to Z_{2m}, the conic α must be an ellipse.

Figure 60: $(Z_{10}, 0)$ Circle System on Ellipse

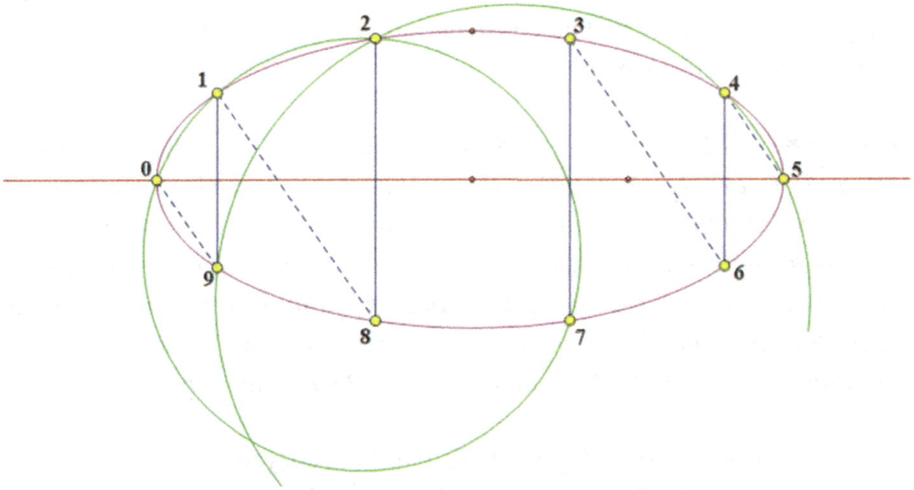

Now consider the auxiliary circles: ((a,b),(c,d), (0,0), (0,0)), ((a,b),(c,d), (0,m),(0,m)), ((a,b),(c,d), (1,m), (1,m)), ((a,b),(c,d), (1,0), (1,0)), where (a,b)+(c,d) = g. These are four distinct circles each through the same two points (a,b), (c,d) and each tangent to the ellipse at a third point. But only two such circles exist: suppose β is a circle through (a,b) and (c,d), then its center O must lie on the perpendicular bisector j of the segment [(a,b),(c,d)]. As O moves along j we find exactly two positions for O where β is tangent to the ellipse. These possibilities are illustrated in Figure 61. We require four such circles but there are only two, so we conclude that no $Z_2 \times Z_{2m}$ circle system, (m ≥ 2) can exist on a noncircular conic.

Figure 61: Two Circles Through points (a,b), (c,d) which are Tangent to the Ellipse

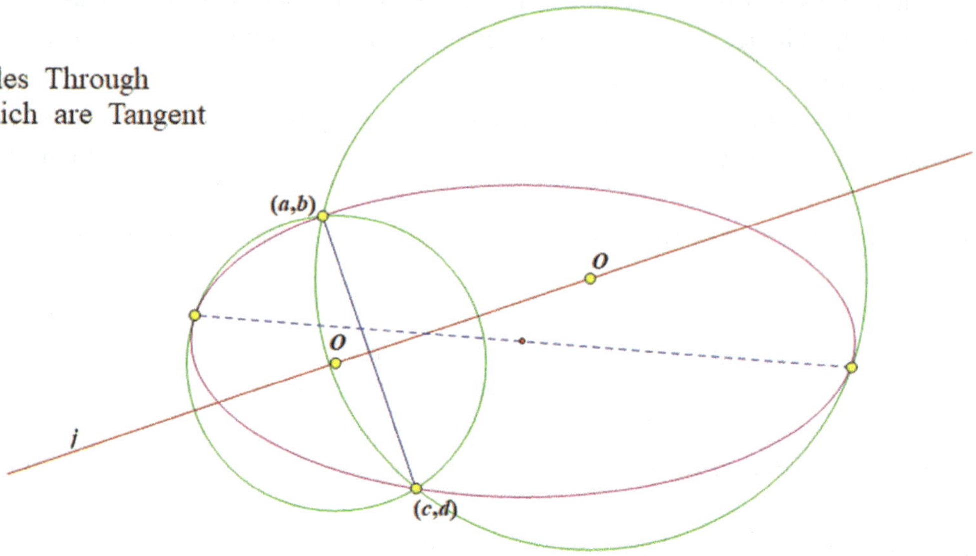

8.5 Direct Product of Circle Systems on a Noncircular Conic

In this Section we show that a direct product construction can be used to create circle systems on a noncircular conic with base group Z × Z. By extension, we can create circle systems with base group Z^n for any positive integer n. The construction is similar to the one given in Theorem 45 for circle systems with the same cubic envelope.

Theorem 88 *Let Ω be a (G,0) circle system, and Ω' a (H,0') circle system on the same noncircular conic α, where 0, 0' are the identities of the abelian groups G, H respectively. We suppose, in accordance with Theorem 81, that the same point P on α and on an axis of symmetry j of α has been labeled with 0 and 0' and that g, -g are reflections across j for every g ∈ G, and similarly h, -h are reflections across j for every h ∈ H. Finally we suppose that Ω and Ω' have only the point P in common. Then a circle system Ω × Ω' with base group G×H and sum (0,0') can be defined on α as follows: (i) relabel each point g ∈ G by (g,0'); (ii) relabel each point h ∈ H by (0,h). Note that the point P will have label (0,0'); (iii) let (g,h) = δ((-g,0'),(0,-h),(0,0')).*

Proof Since the points {(g,h), (-g,0'),(0,-h), (0,0')} are cocyclic, we have by Theorem 80: (12): the lines [(g,h),(0,0')] and [(g,0'),(0,h)] are parallel for every g ∈ G, and h ∈ H. Now suppose s, t ∈ H and s+t = h. Consider the inscribed hexagon H_1 = ((g,0'), (0,h), (0,0'), (g,s), (0,t), (0,s)). We have that lines [(0,h),(0.0')], [(0,t),(0,s)] are parallel by Lemma 82. Also the lines [(0,0'),(g,s)], [(0,s),(g,0')] are parallel by (12). So we can apply Pascal's Theorem (variant) to H_1 to obtain (13) the lines [(g,0'),(0,h)], [(g,s),(0,t)] are parallel. Then from (12), (13) we obtain (14): the lines [(g,h),(0,0')], [(g,s),(0,t)] are parallel. Now let H_2 = ((g,s), (g_0,t), (0,0'), (g,h), (g_0,0'), (0,t)) be a second hexagon inscribed on α where g, g_0 ∈ G. By (12) the lines [(g_0,t),(0,0')], [(g_0,0'),(0,t)] are parallel, and by (14) the lines [(g,h),(0,0')], [(g,s),(0,t)] are parallel. Then applying Pascal's Theorem to H_2 we obtain that the lines [(g,s),(g_0,t)], [(g,h),(g_0,0')] are parallel whenever s+t = h. Consequently we have (15): the set of lines {[(g,s),(g_0,t)] : s+t is a fixed element of H} are mutually parallel. Similarly (16): the set of lines {[(p,h),(q,h_0)] : h,h_0 ∈ H, and p+q is a fixed element of G} are mutually parallel. Now suppose (a,b), (c,d), (u,v), (w,x) ∈ *G × H,* and (a,b) + (c,d) = (u,v) + (w,x) and thus a+c = u+w and b+d = v+x. By (15) the lines [(a,b),(c,d)], [(a,v),(c,x)] are parallel, and by (16) the lines [(a,v),(c,x)], [(u,v),(w,x)] are parallel. We thus obtain that the lines [(a,b),(c,d)], [(u,v),(w,x)] are parallel. We have shown that our construction satisfies the parallelism property stipulated in Lemma 82 for a general 0-sum circle system. It then follows from Theorem 83, that *Ω × Ω'* is a (*G × H,* (0,0')) circle system on α. □

Following the construction given in Theorem 81, we can construct a (Z,0) circle system on a noncircular conic α with zero vertex at a point P on α and on an axis of symmetry of α. In this construction vertex 1 is the generating vertex since its placement determines the positions of the remaining vertices. In principle we can construct a second (Z,0) circle system on α, putting its zero vertex at P and positioning its generating vertex so that the two circle systems have only the zero vertex in common at P. Then using the construction given in Theorem 88 we obtain a (Z × Z, (0,0')) circle system on α. Constructing yet a third (Z,0) circle system on α and choosing the generating vertex so that this new circle system has only the zero vertex in common with the (Z × Z, (0,0')) circle system, we obtain a zero sum circle system on α with base group Z × Z × Z . Repeating this process we can obtain a zero sum circle system on α with base group Z^n for any positive integer n. If α is an ellipse, then we can construct similarly zero sum circle systems on α with base group $Z_m \times Z^n$ for any positive integers m, n. Note that Theorem 88 cannot be used to construct a ($Z_2 \times Z_{2m}$, (0,0)) circle system on an ellipse, (which we have shown to be impossible), since in the construction given in Theorem 88, the (Z_2,0) circle system and the (Z_{2m},0) circle system would have *two* points in common, namely the two points on the ellipse and on the axis of symmetry j. It is not difficult to show that all Z × Z circle systems are equivalent to a Z × Z circle system with sum g where g ∈ {(0,0), (0,1), (0,2), (1,1), (1,2), (1,3), (2,2)} and that all such systems occur as subsystems of a (Z × Z,(0,0)) circle system.

Chapter 9

Circle Chain Conjectures

9.1 The 6-chain Center Line Concurrence on two Lines

In [6] a *circle chain* is defined as follows. Let α, β, c_1, c_2, …, c_n be circles such that for each i, $1 \leq i \leq n-1$ circles c_i and c_{i+1} meet in two points one on α and one on β. The line segment joining these two points is called a *tie.* A *closed circle chain* is a circle chain in which c_n meets c_1 also in two points one on α and one on β. The circles c_i are called *links* and the number of links in a circle chain is the *length* of the chain. The circles α, β are called the *track* and the line segments joining the two contact points of adjacent links are called *ties.* A circle chain is called *regular* in case the pair of ties within any link are disjoint chords, otherwise the circle chain is called *twisted.* In [6] it is shown that any circle chain with odd length can be completed to a closed circle chain by the insertion of a single link. In case the circle chain has length 3, its completion to a closed 4-chain is our Lemma 20 which we have used extensively in Chapter 3 to prove constructions of circle systems. The original conjecture regarding a closed circle chain of length 6 is given in [6] and illustrated in Figure 61.

Conjecture 87 *In a closed circle chain of length 6 with track consisting of two circles, the lines joining the centers of opposite links are concurrent.*

Figure 61: Circle Chain Concurrence of Center Lines for Track Consisting of Two Circles

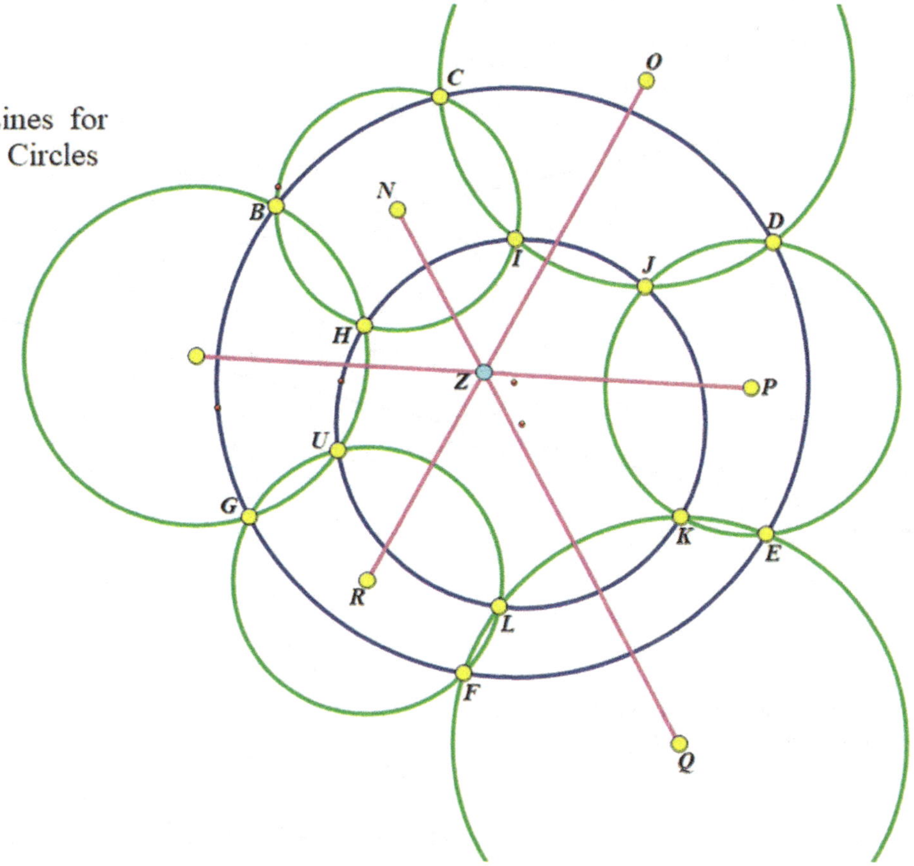

As far as we know Conjecture 87 has never been proved, nor will a proof be given here. Interestingly the concurrence of center lines of a closed 6-chain seems to occur in several settings when the character of the track is modified. Some of these we can prove. We define the track of a circle chain Θ more generally as any curve λ which contains both points of intersection of every pair of adjacent links of Θ. We will start by proving our conjecture in case the track consists of two lines. To do this we require the following two lemmas.

Lemma 88 *Let Θ be a circle chain with track consisting of two lines. Then alternating ties are mutually parallel.*

Proof Suppose first that Θ is regular. In Figure 62 it suffices to show that ∠J = ∠I. Since FBCE and BHGC are cyclic quadrilaterals we have ∠BFE = 180° - ∠BCE and ∠BCG = 180° - ∠BHG. Also ∠BCE + ∠BCG = 180°. Combining we obtain ∠BCE = ∠BHG. Similarly we obtain ∠FEC = 180° - ∠FBC = ∠CBH. Now considering triangles IEC and BJH we have ∠I = 180° - ∠FEC - ∠BCE = 180° - ∠CBH - ∠BHG = ∠J. Now suppose Θ is twisted and consider Figure 63.

Figure 62: Parallelism of ties for Track Consisting of two lines

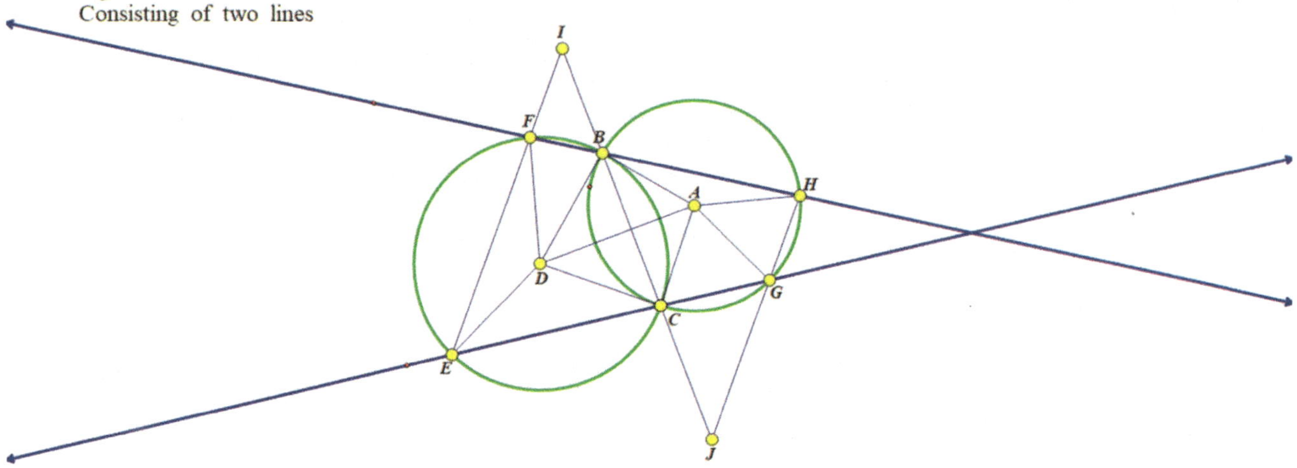

Figure 63: Parallelism property for twisted circle chain on two lines

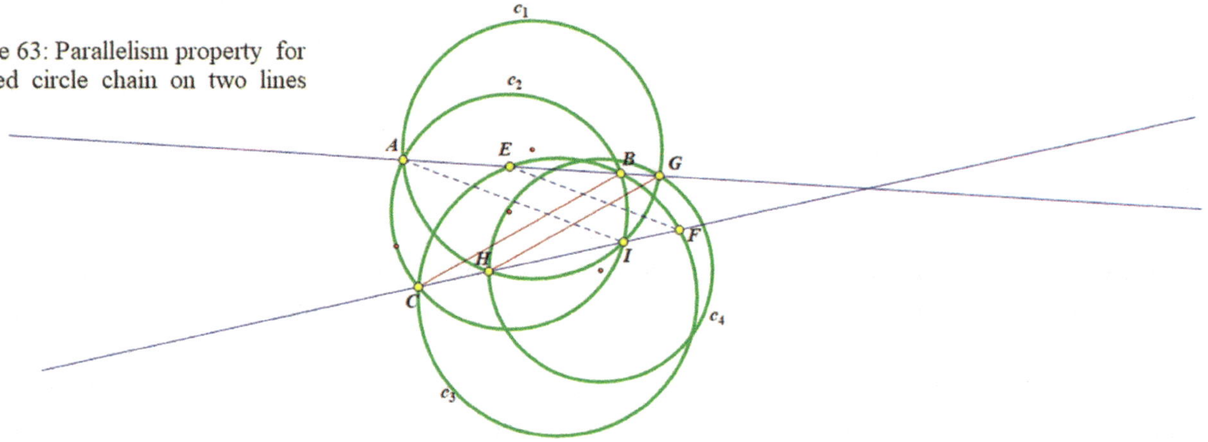

Consider ∠ICB = ∠BAI = ∠GAI = ∠GHI, and thus ties CB, GH ae parallel. Similarly ties AI, EF are parallel.

Lemma 89 *In Figure 64 line segments OT, FI, RL are mutually parallel as are line segments OR, AF, TL. Then lines AL, RF, TL are concurrent.*

Proof Apply a transformation which takes points O,T,L,R to a square say with O = (0,0); T = (0,1); L = (1,1) and R = (1,0). Then we may set A = (0,s); I = (t,1) and F = (t,s) for real numbers s, t. Let S = OI ∩ AL and determine the coordinates of S by intersecting the equations of lines OI and AL. To show F,R,S are collinear just compute the slopes of FR and FS. □

Figure 64: Concurrence Lemma

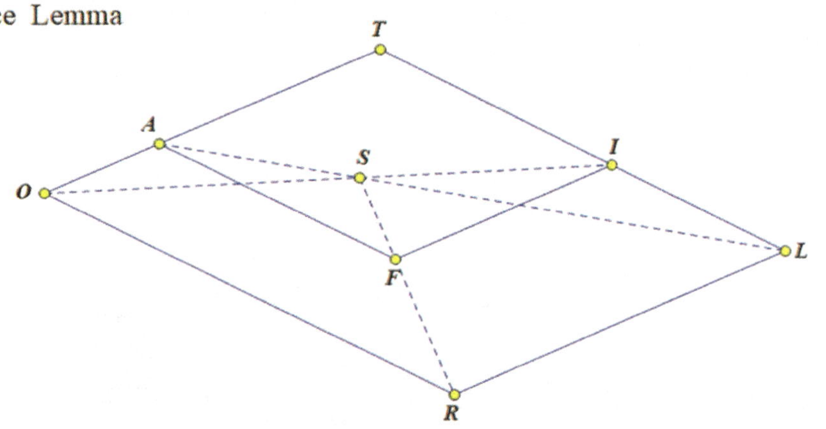

Theorem 90 *Let Θ be a closed 6-chain with track consisting of two lines. Then the center lines of the three opposite links of Θ are concurrent.*

Proof First suppose Θ is regular as in Figure 65. In Figure 65 ties IJ, MN, OR are mutually, as are ties KL, OP, ST by Lemma 88. As a consequence, line segments AF, BC, HE are mutually parallel, as are line segments AH, CD, FE. These six lines form the configuration of Lemma 89, and consequently the center lines FC, AD, BE are concurrent. These are the lines which join the centers of circle pairs $\{c_3,c_6\}$, $\{c_1,c_4\}$ and $\{c_2, c_5\}$ respectively. Now suppose Θ is twisted as in Figure 66. Ties AB, CD, EF ate mutually parallel by Lemma 88, as are ties GH, IJ, KL. As a consequence center lines U_1U_2, U_3U_4, U_5U_6 are mutually parallel as are center lines U_2U_3, U_4U_5, U_6U_1. Now by transforming so that U_2,U_3,U_4 form the corners of a rectangle we obtain a figure similar to Figure 64 when the desired concurrence follows from Lemma 89. □

Figure 65: Regular 6-chain center line concurrence for track consisting of two lines

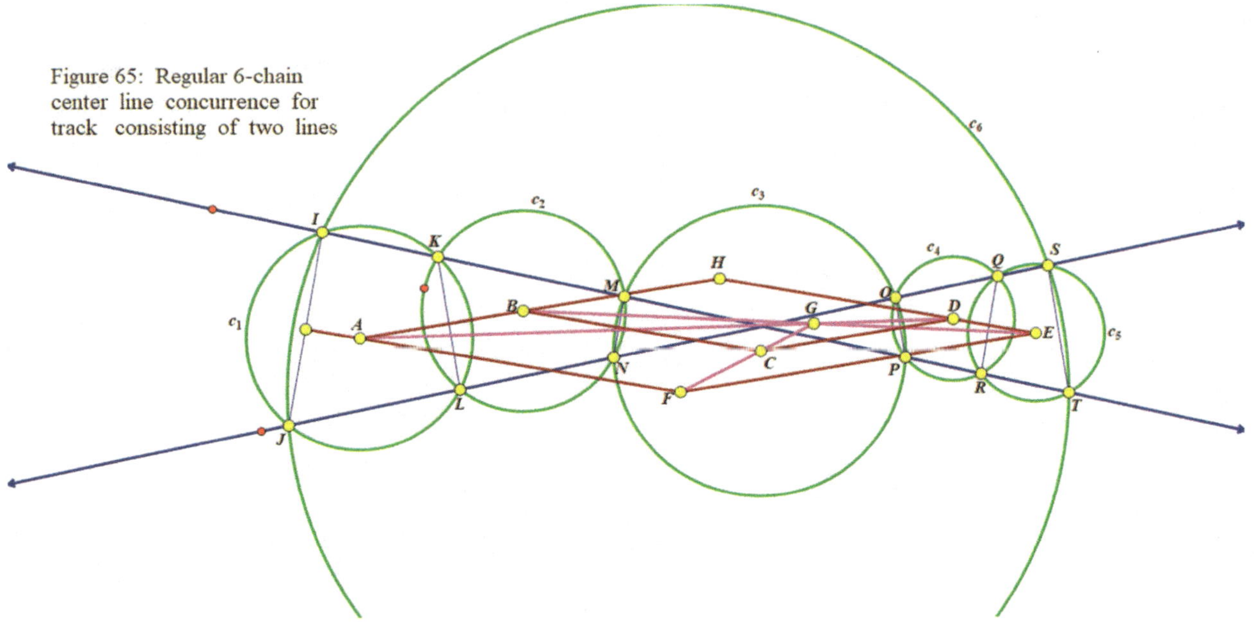

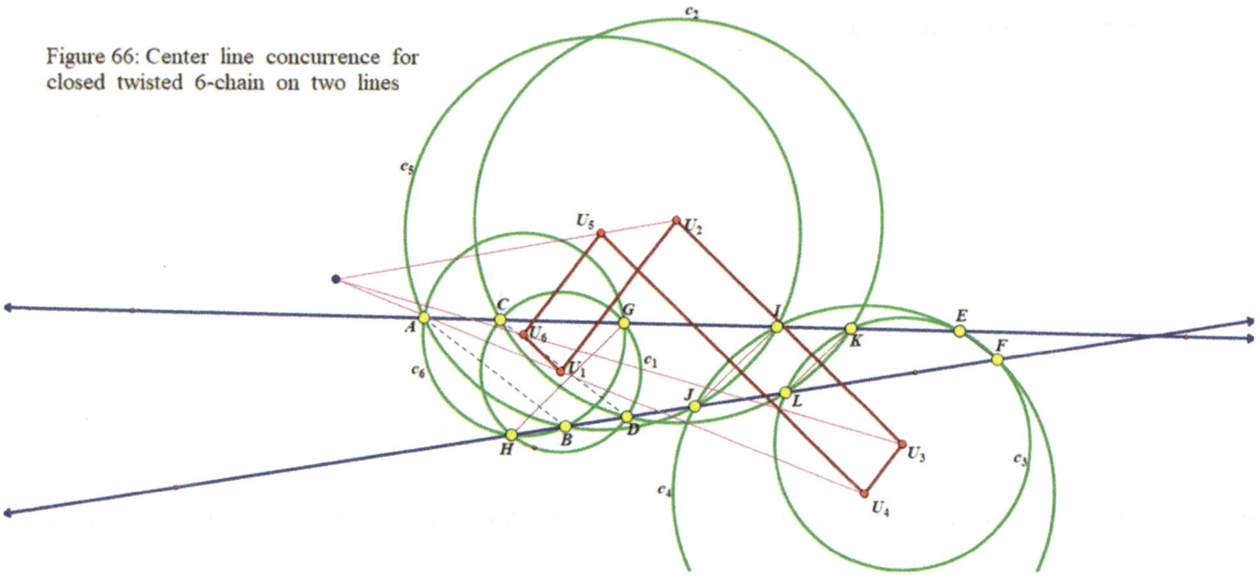

Figure 66: Center line concurrence for closed twisted 6-chain on two lines

In Theorem 90 the existence of circle c_6 should be proved. In fact we can show the following.

Theorem 91 *Let Θ be a circle chain with track consisting of two lines. Then the four endpoints belonging to any two nonparallel ties are cocyclic.*

Proof Suppose first that Θ is regular and consider Figure 67. In this Figure ∠O meets the track lines in four cocyclic points V,W,U,X and and ∠T meets the track lines in points A,B,C,D. Suppose also that the sides of ∠T are parallel to the sides of ∠O. The corresponding angles ∠UVW, ∠DAB are congruent as are the corresponding angles ∠UXW, ∠DCB. As UVWX is a cyclic quadrilateral we have ∠UVW + ∠UXW = 180° and thus ∠DAB + ∠DCB = 180° with the consequence that DABC is also a cyclic quadrilateral. Now due to the parallelism property of Lemma 88 the angles made by two nonparallel ties are all congruent and corresponding sides are parallel. The Theorem then follows from the first part of the proof. Now suppose Θ is twisted and consider Figure 68.

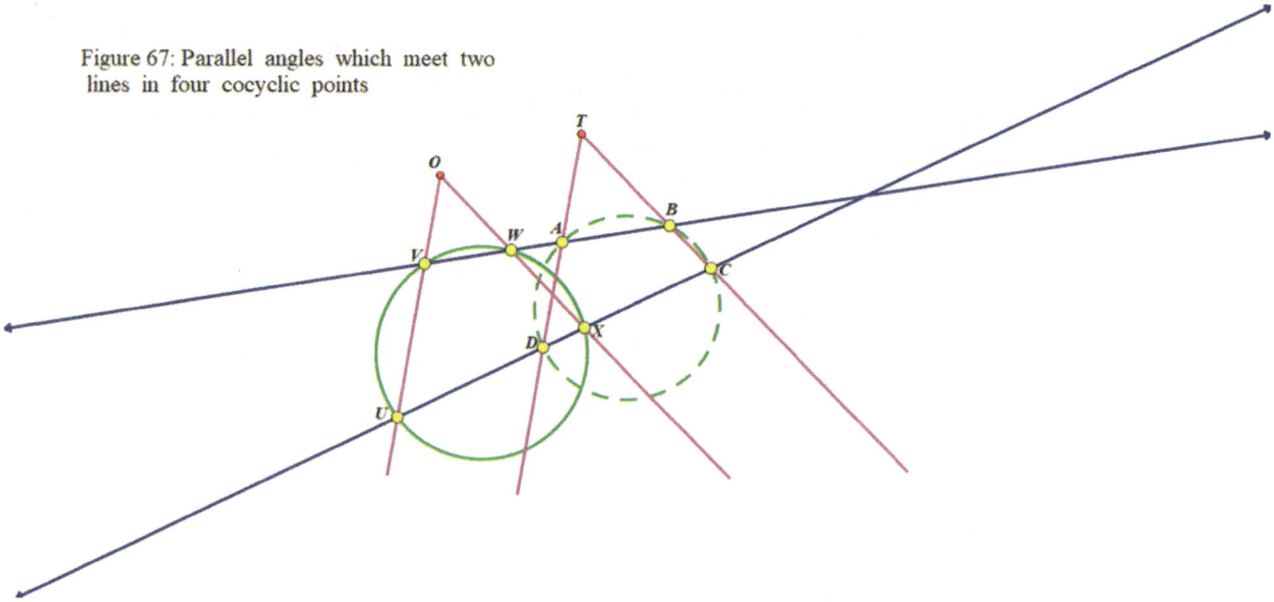

Figure 67: Parallel angles which meet two lines in four cocyclic points

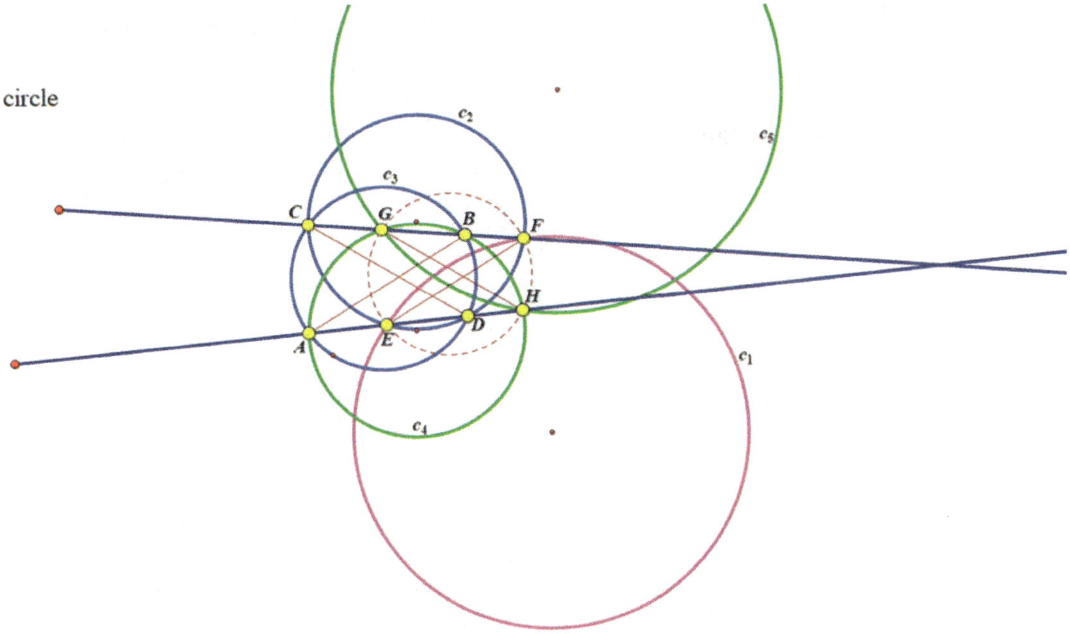

Figure 68: Twisted circle chain on two lines

In Figure 68, GH, EF are nonparallel ties and we want to show that points E,G,F,H are cocyclic. A twist occurs in circle c_3 and by Lemma 88 we know that CD is parallel to GH and AB is parallel to EF. The essence of what needs to be shown is illustrated in Figure 69 where we have the circle ACBD and then just put line segments GH parallel to CD and EF parallel to AB. Note that ∠BCD = ∠FGH = ∠FEH = 180° - ∠DEF and thus CDEF is a cyclic quadrilateral. Now invert Figure 69 by a circle centered at E to obtain Figure 70 where circle CDEF is inverted to line D'C'F'. To show that E,G,F,H are cocyclic it suffices to show that points G', F', H' are collinear. Consider: ∠B'G'F' = 180° - ∠B'C'F' = ∠B'C'D' = ∠B'A'D' = 180° - ∠B'A'H', and considering inscribed quadrilateral G'B'A'H' we have ∠B'G'H' = 180° - ∠B'A'H'. Thus ∠B'G'F' = ∠B'G'H' with the consequence that points G', F', H' are collinear.

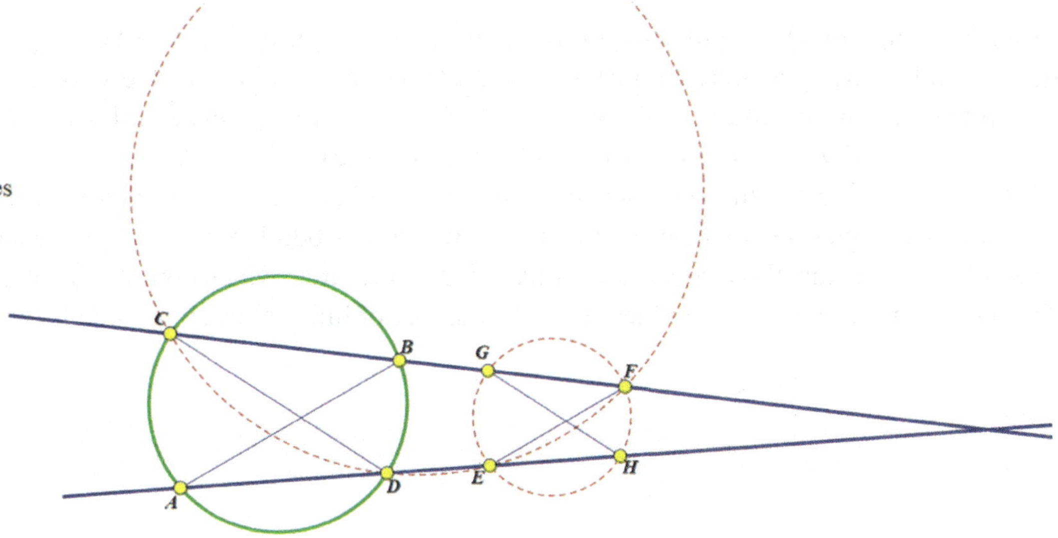

Figure 69: parallel crosses on two lines

Figure 70: Inverse of Figure 69

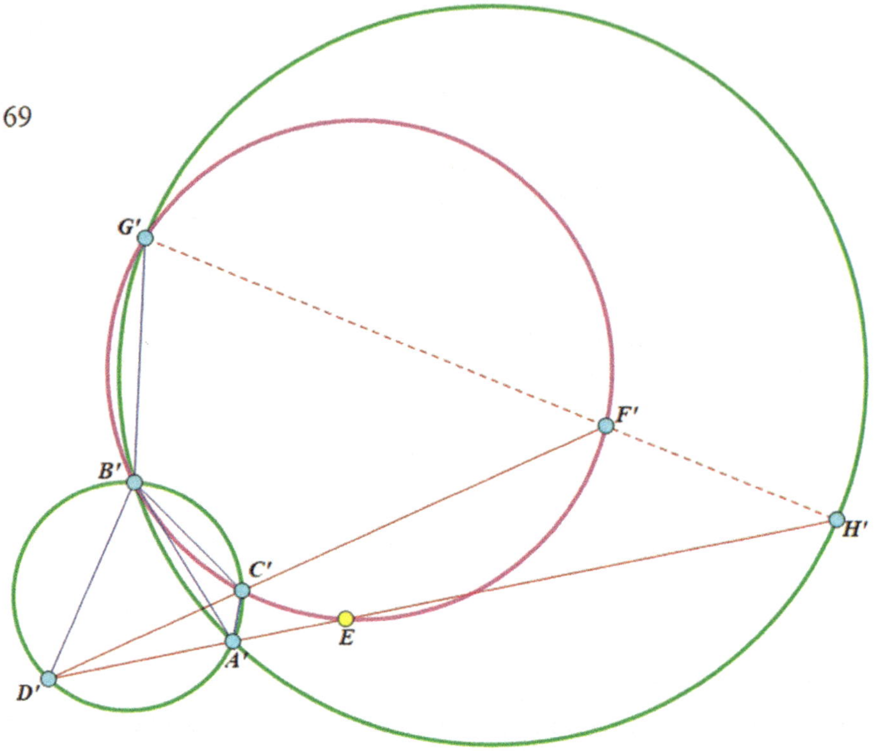

As a consequence of Theorem 91, any odd length circle chain with track consisting of two lines can be extended to a closed (even) length circle chain by the insertion of a single link. In case the track consists of two circles as in Conjecture 87, the same result is proved in [6]. Here we present a much simpler proof.

Theorem 92 *Let Θ be a circle chain with odd length and with track consisting of two circles α,β. Then Θ can be closed by the addition of a single circle.*

Proof First suppose Θ is regular with links c_1, c_2, c_3 as in Figure 71. Let M be the point at the junction of α, c_1, c_2 and invert by a circle with center at M. Then α, c_1, c_2 are converted into lines as in Figure 72. In the inverted Figure it suffices to show that A'B'G'F' is a cyclic quadrilateral. Consider: ∠B'G'F' = 360° - ∠B'G'D' - ∠F'G'D' = 360° - (180° - ∠C') - (180° - ∠E') = ∠C' + ∠E'. Also ∠A' = 180° - ∠C'- ∠E', and so ∠B'G'F' + ∠A' = 180° with the consequence that A'B'G'F' is a cyclic quadrilateral. It follows that the circle c_4 = ABGF completes Θ to a closed 4-chain. Now if Θ has odd length ≥ 5 then we can shorten Θ by two links by replacing any three consecutive links of Θ by the circle which completes these to a closed 4-chain. The Theorem then follows by induction on the number of links. Now suppose Θ is twisted as in Figure 73.

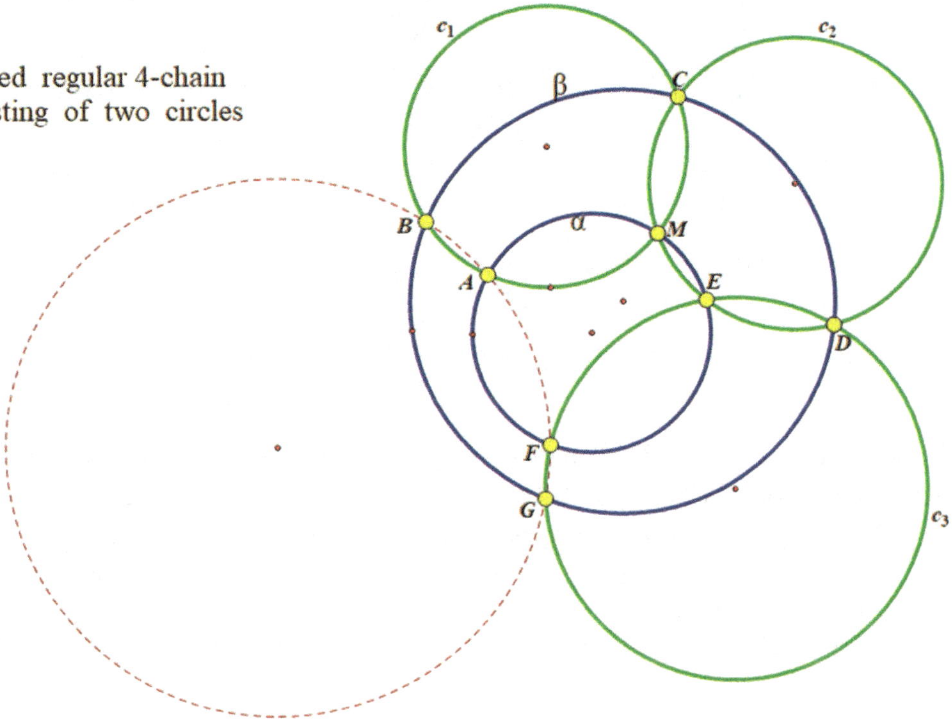

Figure 71: Closed regular 4-chain on track consisting of two circles

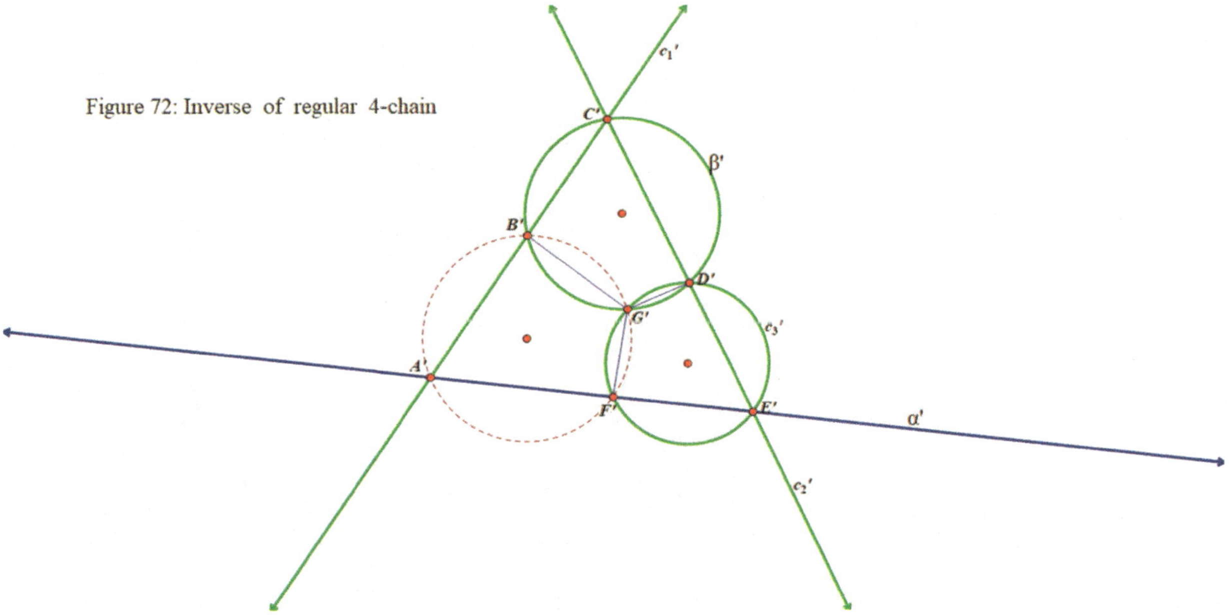

Figure 72: Inverse of regular 4-chain

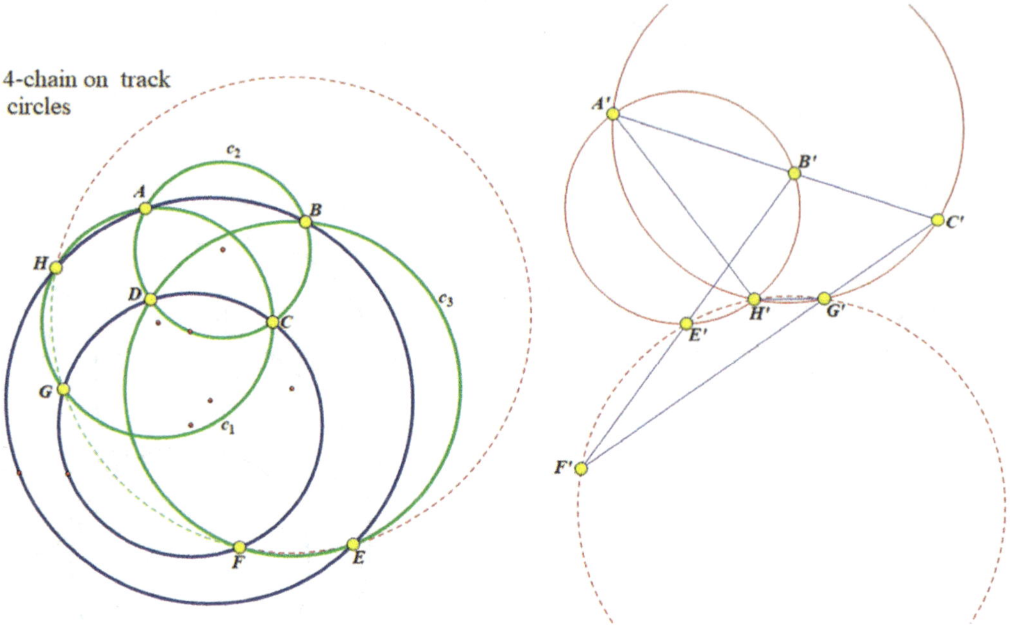

Figure 73: Twisted 4-chain on track consisting of two circles

We will show that the twisted 3-chain consisting of circles c_1, c_2, c_3 can be extended to a closed 4-chain by demonstrating the cocyclicity of points E,F,G,H. Invert by a circle centered at point D to obtain the primed figure at right in Figure 73. It then suffices to show that points E',F',G',H' are cocyclic. Consider: ∠F'E'H' = 180° - ∠B'E'H' = 180° - ∠B'A'H' = ∠H'G'C' = 180° - ∠H'G'F', from which it follows that F'E'H'G' is a cyclic quadrilateral. It follows as in the part of the argument for a regular circle chain, that any odd length twisted circle chain can be extended to a closed circle chain by the addition of a single link. □

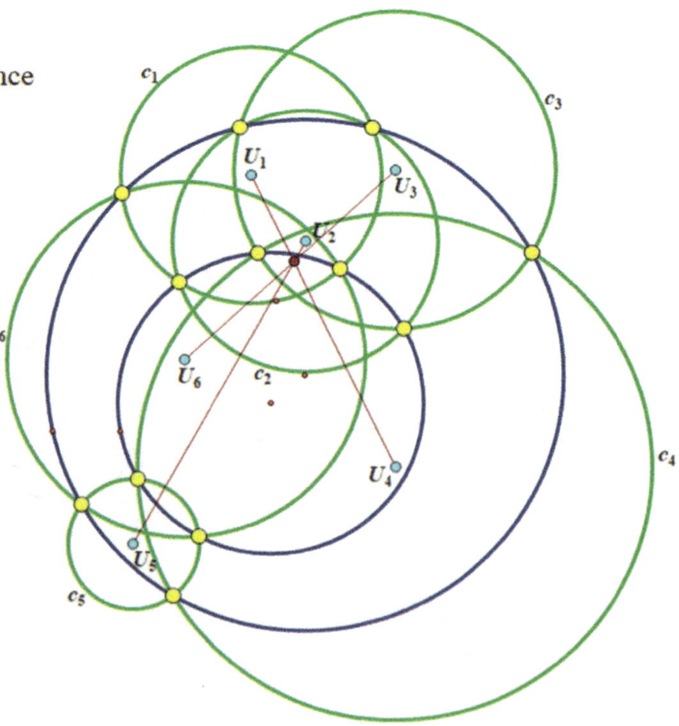

Figure 74: Center line concurrence for closed twisted 6-chain on track consisting of two circles

Conjecture 87 also evidently holds in case Θ is a closed twisted 6-chain on two circles as in Figure 74.

9.2 The 6-chain Center Line Concurrence on a Conic

In Chapter 8 in Theorem 83 it was shown that ternary hypercommutativity holds on any noncircular conic, i.e., if δ(a,b,c) is defined to be the fourth point of contact of circle (a,b,c) with a noncircular conic α, then (α,δ) is a THA. In particular the expression δ(δ(a,b,c), δ(d,e,f), δ(g,h,k)) is invariant under any permutation of its nine variables. Our first result of this Section is an analog of Theorem 92 which holds not only for noncircular conics but for any curve upon which a THA can be defined.

Theorem 93 *Let λ be any curve upon which a THA can be defined, ie., any circle which meets λ in three points a,b,c meets the curve also in a unique fourth point δ(a,b,c) counting multiplicities, and for which (λ,δ) is a THA. Then any odd length circle chain with λ as track can be extended by the addition of a single link, to a closed circle chain on λ.*

Proof Consider the circle chain (c_1, c_2, c_3) of Figure 75. To complete this 3-chain to a closed 4-chain we must show that points G,H,A,B are cocyclic. Consider: δ(G,H,A) = δ(δ(E,F,H), H, δ(B,C,D)) = δ(δ(δ(C,D,F),F,H), H, δ(B,C,D))= δ(δ(B,F,H), H, δ(δ(C,D,F),C,D))= δ(δ(B,F,H), H,F) = B, and thus points G,H,A,B are indeed cocyclic. The remainder of the argument is the same as the one used in Theorem 92. □

Figure 75: Closed 6-chain on parabola

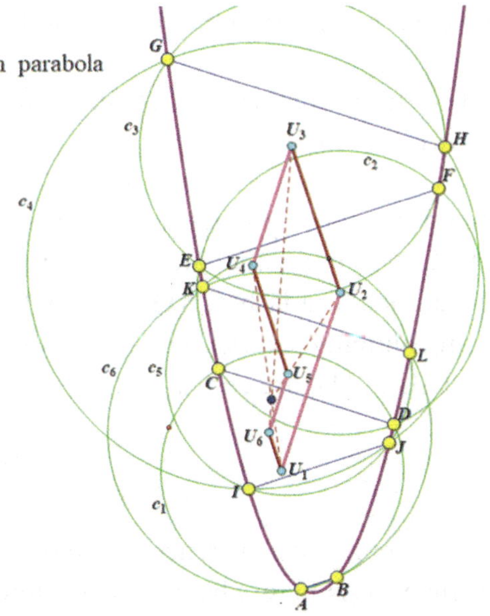

Next we have an analog of the parallelism property given in Lemma 88.

Theorem 94 *Let Θ be a circle chain with track consisting of a noncircular conic. Then alternating ties of Θ are mutually parallel.*

Proof In Figure 76 we must show that ties [A,B] and [C,D] are parallel. By Lemma 82, [A,B] = [δ(E,F,B), δ(A,G,H)] is parallel to [E.I] = [δ(A,F,B), δ(E,G,H)]. Also by Lemma 82 [C,D] = [δ(G,H,D), δ(C,E,F)] is parallel to [E,I] = [δ(C,D,F), δ(G,H,E)], and thus [A,B] is parallel to [C,D]. □

Figure 76: Parallelism property for circle chain on conic

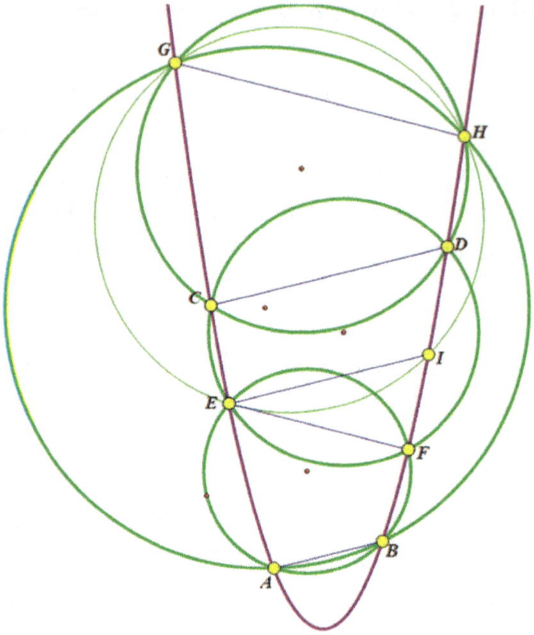

With only these preliminaries we can prove our main Conjecture for a closed 6-chain on a noncircular conic.

Theorem 95 *Let Θ be a closed 6-chain with track consisting of a noncircular conic α. Then the three lines joining the centers of opposite links of Θ are concurrent.*

Proof By Theorem 93 any open 5-chain on α can be completed to a closed 6-chain on α as in Figure 75. With reference to this Figure, ties AB, IJ, EF are mutually parallel by Theorem 94, with the consequence that lines U_2U_3, U_4U_5, U_6U_1 are mutually parallel. Similarly ties GH, KL, CD are mutually parallel with the consequence that lines U_1U_2, U_3U_4, U_5U_6 are mutually parallel. The center points then form the configuration of Lemma 89 and thus center lines U_1U_4, U_2U_5, U_6U_3 are concurrent. □

9.3 Conjectures

Consider once again Conjecture 87 for the center line concurrence for a closed 6-chain with track consisting of two circles as illustrated in Figure 61. Invert this Figure by a circle centered at vertex I to obtain Figure 77. Circles c_1, c_2 and one

Figure 77: Inverse of Figure 61

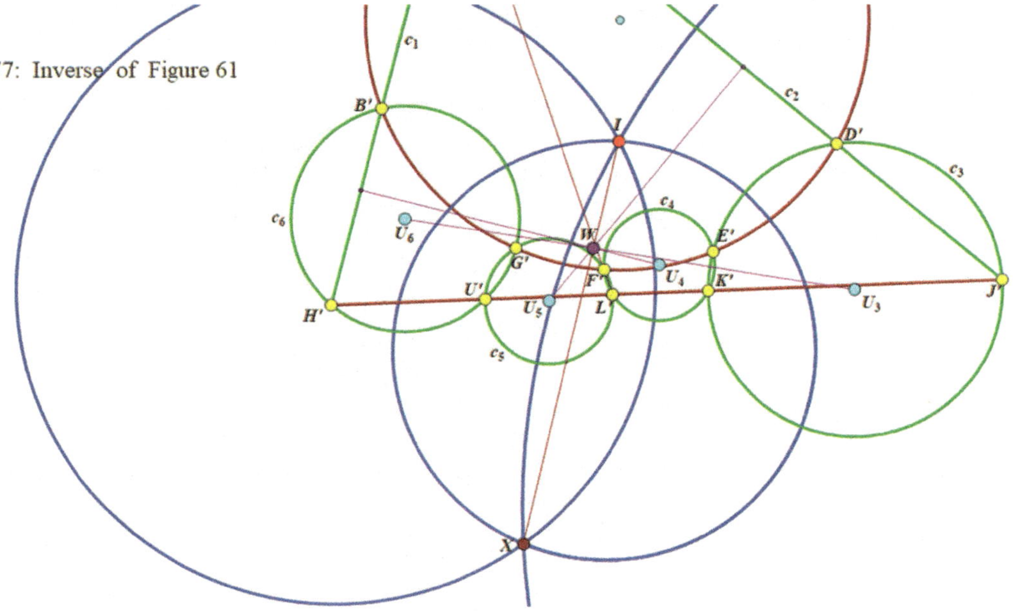

of the track circles have been changed to lines and the center lines joining the centers of opposite circles in the 6-chain have been changed into circles through the same two points I, X and which are orthogonal to the circles whose centers they join. Our Conjecture 87 is proved if it can be shown that these three circles indeed contain the common point X . Interestingly, our center line concurrence also seems to hold in the inverted version of Figure 77. We interpret the line joining the *centers* of c_2 and c_5 as the line through U_5 and perpendicular to the line c_2, and similarly the line through U_4 and perpendicular to the line c_1 as the line joining the centers of c_1 and c_4. We then find that these two lines along with the center line U_3U_6 are concurrent. Moreover, if W denotes the point of concurrence, then points I, W, X are concurrent. We leave these observations as conjectures. The concurrence at W should be provable by coordinatizing the points in Figure 77 and working with equations of lines and circles.

In the following conjecture we maintain that our center line concurrence holds also in case the track consists of a self-inversive cubic curve or the inverse of such a curve.

Conjecture 96 *Let Θ be a closed 6-chain whose track consists of a self-inversive cubic curve or the inverse of such a curve. Then the lines joining the centers of the three pairs of opposite links of Θ are concurrent.*

In Figure 78 we illustrate the center line concurrence for a track consisting of a self-inversive cubic curve and in Figure 79 the same concurrence is illustrated for a track consisting of a bicircular quartic obtained as an inverse of a self-inversive cubic.

Figure 78: Closed 6-chain on Type IV cubic

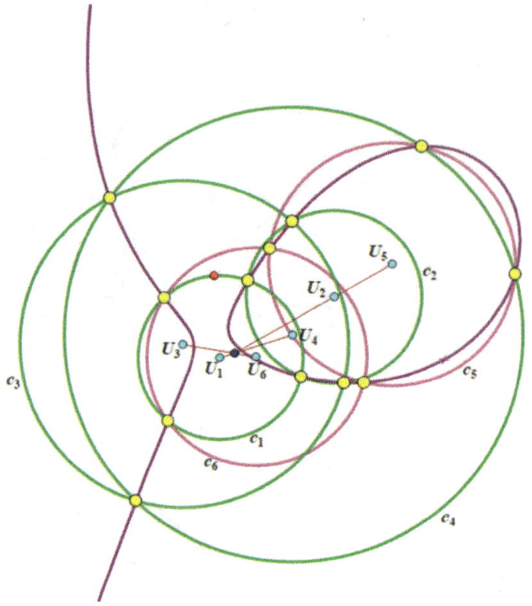

Figure 79: Closed 6-chain on bicircular quartic curve

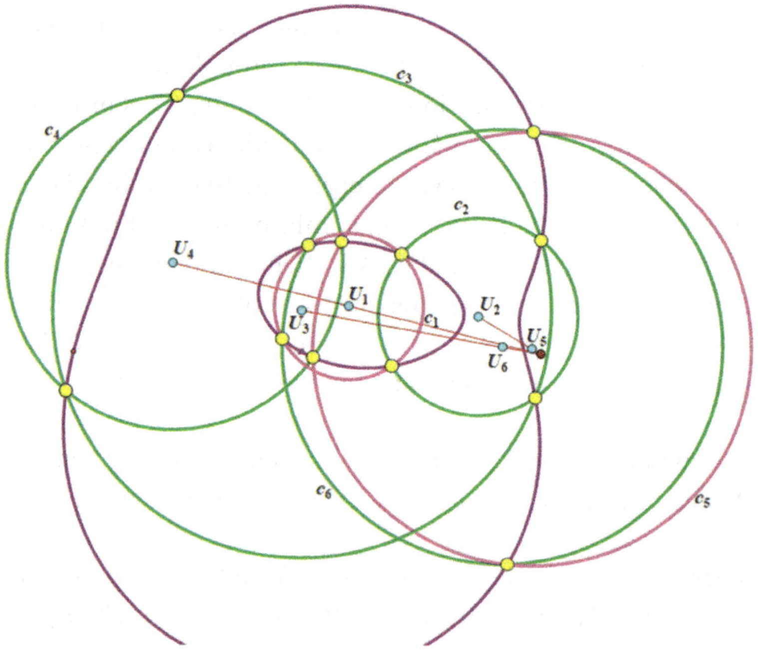

Closed 6-chains occur in abundance in circle systems with sufficiently many vertices. For example, in a $(Z_{14},0)$ circle system there are 15 balanced circles each formed by the points in any two of the pairs {1,13}, {2,12}, {3,11}, {4,10}, {5,9}, {6,8}. These circles form numerous closed 6-chains and thus there are numerous triples of concurrent center lines. Such 6-chains can be described as *strong* since the endpoints of *any* two ties are cocyclic. There also exist in the $(Z_{14},0)$ 6-chains consisting of circles of the form (a,b,c,d) where c+d = -(a+b), for example {(0,1,6,7), (6,7,13,2), (13,2,9,4), (9,4,12,3), (12,3,5,8), (5,8,0,1)}. These give rise to even more concurrent center lines.

Chapter 10

Circle System Parameters

10.1 Sizes of Circle Systems with Base Z_n

Let Ω be a (G,g) circle system with G finite. The *size* q of Ω is the number of circles in Ω. If $x \in G$ then $\delta(x)$, the *degree of x in Ω,* denotes the number of circles in Ω which contain x. Summing degrees counts every circle in Ω four times, so we have:

$$(1)\ q = (1/4)\textstyle\sum_{x\in G}\delta(x).$$

So to determine q we need a formula for $\delta(x)$. For $x \in G$ let $\tau(x)$ denote the number of three element subsets of G which sum to x and let $\rho(x)$ denote the number of two element subsets of G which sum to x. Then we have:

$$(2)\ \delta(x) = \tau(g\text{-}x) - \rho(g - 2x) + \varepsilon$$

Where $\varepsilon = 0$ if $4x = g$ and $\varepsilon = 1$ otherwise. The explanation of (2) is as follows. A circle in Ω which contains x consists of x and a three element set which sums to g-x. The expression $\tau(g\text{-}x)$ counts all three element sets which sum to g-x, but among these we need to exclude those which contain x. There are $\rho(g - 2x)$ two element subsets which sum to $g - 2x$ so this counts the number of excluded sets. This is precisely the number of sets to be excluded in case $4x = g$, but if $4x \neq g$ then we have excluded one too many since there is a two element set which sums to $g - 2x$ and which contains x. In this case we take $\varepsilon = 1$.

For $x \in G$ let $\xi(x) = |g \in G : g+g = x\}|$. Then:

$$(3)\ \rho(x) = (1/2)(|G| - \xi(x))$$

The explanation of (3) is as follows. In the multiplication table for G, the element x occurs exactly once in each row and column. The expression $\xi(x)$ counts the number of times x occurs on the main diagonal of the table. So x occupies $|G| - \xi(x)$ nondiagonal locations. Each pair of unequal elements a,b which sum to x yield two x entries, one in row a and column b and one in row b and column a. Thus there are $(1/2)(|G| - \xi(x))$ two element sets {a,b} which sum to x.

Now suppose $G = Z_n$. Then we have the following:

Theorem 99 *Let Ω be a circle system with base Z_n and suppose gcd(n,4) = 1. Then Ω is equivalent to a $(Z_n, 0)$ circle system and has size $q = (1/n)\binom{n}{4} = (1/24)(n\text{-}1)(n\text{-}2)(n\text{-}3)$.*

Proof Since gcd(n,4) = 1, $Z_n = \langle 4\rangle$. Let q_i denote the number of four element subsets of Z_n which sum to i. If S = {a,b,c,d} is a four element set which sums to 0, then by adding 1 to each member of S we obtain a four element subset of Z_n which sums to 4. Thus $q_0 = q_4$. But since gcd(n,4) = 1, $Z_n = \langle 4\rangle$ and thus we obtain $q_0 = q_i$ for each $i \in Z_n$. This yields $nq_0 = \binom{n}{4}$. □

Now suppose that $n = 2m$ is even. In this case $\xi(x) = 0$ if x is odd, and $\xi(x) = 2$ if x is even. Note that if $x = 2s$, then $\{s, s+m\}$ constitute precisely the set of elements in Z_{2m} which sum with themselves to x. Then for $G = Z_{2m}$ equation (3) yields:

(4) $\rho(x) = n/2 = m$ *if x is odd;* $\rho(x) = m - 1$ *if x is even.* $(G = Z_{2m})$

Let $G = Z_{2m}$ and consider that $\tau(x)$ is precisely the size of a (G, x) triple system. The sizes and equivalence classes of triple systems with base Z_n are determined in [2]:

Theorem 100 *Let T be a geometric triple system with base Z_n. If $gcd(n,3) = 1$ then T is equivalent to a $(Z_n, 0)$ triple system and has size* $(1/n)\binom{n}{3} = (1/6)(n-1)(n-2)$. □

Theorem 101 *Let T be a geometric triple system with base Z_n. If $3|n$ then T is equivalent to either a $(Z_n, 0)$ or a $(Z_n, 1)$ triple system. In the first case the size of T is $(1/6)(n^2 - 3n + 6)$, and in the second case the size of T is $(1/6)(n^2 - 3n)$.* □

Now with regard to circle systems we have:

Theorem 102 *Let Ω be a circle system with base Z_n. (i) If $gcd(n,4) = 1$, then Ω is equivalent to a $(Z_n, 0)$ circle system and has size $(1/24)(n-1)(n-2)(n-3)$.(ii) If $gcd(n,4) = 2$, then Ω is equivalent to either a $(Z_n,0)$ or $(Z_n, 1)$ circle system. If $gcd(n,4) = 4$ then Ω is equivalent to a $(Z_n,0)$, $(Z_n, 1)$ or $(Z_n, 2)$ circle system.* □

We will determine here the sizes for the circle systems in (ii) and (iii). Let Ω be a $(Z_n,0)$ circle system with $n = 4k+2 = 2m$ and suppose $gcd(n,3) = 1$. In formula (2) $\tau(g - x)$ is the size of a $(Z_n, g-x)$ triple system. Then by Theorem 100 we have $\tau(g - x) = (1/6)(n-1)(n-2)$. Then from (2), (4) we obtain: $\delta(x) = (1/6)(n-1)(n-2) - \rho(-2x) + \varepsilon = (1/6)(n-1)(n-2) - (m - 1) + \varepsilon$, where we take $\varepsilon = 0$ if $4x = 0$ and otherwise $\varepsilon = 1$. In this case $\{0,m\}$ are the only two elements in Z_n which satisfy $4x = 0$, so we have:

$$\delta(0) = \delta(m) = (1/6)(n-1)(n-2) - ((n/2) - 1) = (1/6)(n-2)(n-4) \text{ and:}$$

$$\delta(x) = (1,6)(n-1)(n-2) - ((n/2)-1) + 1 = (1/6)(n^2 - 6n + 14) \text{ for } k \,\varepsilon\, Z_n, k \neq 0, m.$$

Then from (1) we obtain:

$$q = (1/4)(\delta(0) + \delta(m) + \textstyle\sum_{x \in Z_n, x \neq 0,m} \delta(x)) =$$

$$(1/4)\,((1/3)(n-2)(n-4) + (n-2)(1/6)(n^2 - 6n + 14))\,.$$

$$q = (1/24)(n-2)(n^2 - 4n + 6).$$

Conclusion: If Ω is a $(Z_n, 0)$ circle system and $gcd(n,4) = 2$, $gcd(n,3) = 1$, then the size of Ω is given by $q = (1/24)(n-2)(n^2 - 4n + 6)$.

Now suppose Ω is a $(Z_n, 0)$ circle system with $n = 4k+2 = 2m$ and such that $3|n$. From (2) we have:

(5) $\delta(x) = \tau(-x) - \rho(-2x) + \varepsilon$ *, where* $\varepsilon = 0$ *if* $4x = 0$ *and* $\varepsilon = 1$ *otherwise.*

Since -2x is even, by (4) $\rho(-2x) = m - 1 = \frac{n-2}{2}$ for every $x \in Z_n$. Consider τ-(x) is the size of a $(Z_n, -x)$ triple system T. If x is a multiple of 3, then so is -x, and T is equivalent to a $(Z_n, 0)$ triple system which has size $(1/6)(n^2 - 3n + 6)$ by Theorem 101. Otherwise T is equivalent to a $(Z_n, 1)$ triple system and has size $(1/6)(n^2 - 3n)$ by Theorem 101. The only elements in Z_n which satisfy $4x = 0$ are {0,m} so (5) yields:

$$\delta(x) = (1/6)(n^2 - 3n + 6) - (1/2)(n-2) = (1/6)(n^2 - 6n + 12) \text{ if } x = 0,m.$$

Note that m is a multiple of 3. Consider: $\delta(x) = (1/6)(n^2 - 6n + 12) + 1$ if x is a multiple of 3 different from 0,m and $\delta(x) = (1/6)(n^2 - 3n) - (1/2)(n-2) + 1 = (1/6)(n^2 - 6n + 12)$ if x is not a multiple of 3. We then obtain:

$$\sum_{x \in Z_n} \delta(x) = (\frac{n}{3} - 2)(\frac{n^2-6n+12}{6} + 1) + (\frac{2n}{3} + 2)(\frac{n^2-6n+12}{6}) = \frac{n^3-6n^2+14n-12}{6}.$$

So from (1) we obtain $q = \frac{n^3-6n^2+14n-12}{24} = \frac{(n-2)(n^2-4n+6)}{24}$. This is the same count as in the previous case so we conclude:

Theorem 103 *If* Ω *is a* $(Z_n, 0)$ *circle system with* $\gcd(n,4) = 2$ *, then the size of* Ω *is given by* $q = \frac{(n-2)(n^2-4n+6)}{24}$. □

Now suppose Ω is a $(Z_n, 1)$ circle system with n even and such that $3|n$. From (2) we obtain $\delta(x) = \tau(1-x) - \rho(1 - 2x) + \varepsilon$, where $\varepsilon = 0$ if $4x = 1$, and $\varepsilon = 1$ otherwise. Since $4x = 1$ has no solutions in Z_n, we have $\varepsilon = 1$ for all x. Also $1 - 2x$ is always odd and so $\rho(1 - 2x) = n/2$ for all x by (4) thus:

(6) $\delta(x) = \tau(1\text{-}x) - (n/2) + 1$ *for all x in* Z_n

Consider: $\tau(x)$ is the size of a (Z_n, x) triple system T. By Theorem 101, T is equivalent to either a $(Z_n, 0)$ or a $(Z_n, 1)$ triple system. In the first case T has size $(1/6)(n^2 - 3n + 6)$ and in the second case T has size $(1/6)(n^2 - 3n)$. If x is a multiple of 3 then T is equivalent to a $((Z_n, 0)$ triple system, and otherwise T is equivalent to a $(Z_n, 1)$ triple system. Thus:

$$\delta(x) = (1/6)(n^2 - 3n + 6) - (n/2) + 1 = (1/6)(n^2 - 6n + 12) \text{ if } 3|(1-x);$$

$$\delta(x) = (1/6)(n^2 - 3n) - (n/2) + 1 = (1/6)(n^2 - 6n + 6) \text{ otherwise.}$$

In Z_n there are n/3 elements x for which 1-x is a multiple of 3, and so:

$$\sum_{x \in Z_n} \delta(x) = (n/3)(1/6)(n^2 - 6n + 12) + (2n/3)(1/6)(n^2 - 6n + 6) = \frac{n(n-2)(n-4)}{6}$$

So from (1) we obtain $q = \frac{n(n-2)(n-4)}{24}$.

Conclusion: If Ω is a $(Z_n, 1)$ circle system with n even and such that $3|n$, then the size of Ω is given by $q = \frac{n(n-2)(n-4)}{24}$.

Now suppse Ω is a $(Z_n, 0)$ circle system with gcd(n,4) = 4 and gcd(n,3) = 1. Then (2) yields $\delta(x) = \tau(-x) - \rho(-2x) + \varepsilon$. If n = 4t then the equation 4x = 0 has solutions {0,t,2t,3t} in Z_n. Also by(4) $\rho(-2x) = \frac{n-2}{2}$ for all x since -2x is even. The expression $\tau(x)$ gives the size of a (Z_n, x) triple system T. Since gcd(n,3) = 1, T is equivalent to $(Z_n, 0)$ triple system and has size (1/6)(n-1)(n-2) by Theorem 100. Thus:

$$\delta(x) = (1/6)(n-1)(n-2) - (1/2)(n-2) = (1/6)(n-2)(n-4) \text{ if } x \in \{0,t,2t,3t\}, \text{ and}$$

$$\delta(x) = (1/6)(n-1)(n-2) - (1/2)(n-2) + 1 = (1/6)(n-2)(n-4) + 1 = (1/6)(n^2 - 6n + 14) \text{ otherwise.}$$

Thus:

$\sum_{x \in Z_n} \delta(x) = 4(1/6)(n-2)(n-4) + (n-4)(1/6)(n^2 - 6n + 14) = \frac{(n-4)(n^2-2n+6)}{6}$, and so by (1) we have: $q = \frac{(n-4)(n^2-2n+6)}{24}$.

Conclusion: If Ω is a $(Z_n, 0)$ circle system with gcd(n,4) = 4 and gcd(n,3) = 1, then the size of Ω is $q = \frac{(n-4)(n^2-2n+6)}{24}$.

Now suppose Ω is a $(Z_n, 0)$ with gcd(n,4) = 4 and such that $3|n$. Then (2) yields:

$\delta(x) = \tau(-x) - \rho(-2x) + \varepsilon$. If n = 4t, then 4x = 0 implies $x \in \{0,t,2t,3t\}$. Also $\rho(-2x) = (1/2)(n-2)$ for all x by (4) since -2x is even. If $3|x$ and thus $3|(-x)$, $\tau(-x) = (1/6)(n^2 - 3n + 6)$ by Theorem 3. Otherwise $\tau(-x) = (1/6)(n^2 - 3n)$ by Theorem 3. Consequently $\delta(x) = (1/6)(n^2 - 3n + 6) - (1/2)(n-2) = (1/6)(n^2 - 6n + 12)$ if $x \in \{0,t,2t,3t\}$. Note that $3|x$ in this case;

$$\delta(x) = (1/6)(n^2 - 3n + 6) - (1/2)(n-2) + 1 = (1/6)(n^2 - 6n + 18) \text{ if } 3|x \text{ and } x \notin \{0,t,2t,3t\};$$

$$\delta(x) = (1/6)(n^2 - 3n) - (1/2)(n-2) + 1 = (1/6)(n^2 - 6n + 12) \text{ if } \gcd(3,x) = 1.$$

We then obtain:

$\sum_{x \in Z_n} \delta(x) = 4(1/6)(n^2 - 6n + 12) + (\frac{n}{3} - 4)(1/6)(n^2 - 6n + 18) + (\frac{2n}{3})(1/6)(n^2 - 6n + 12) = \frac{(n-4)(n^2-2n+6)}{6}$. So by (1) we have $q = \frac{(n-4)(n^2-2n+6)}{24}$. Note this is the same count as in the previous case so we have

Theorem 104 *If Ω is a $(Z_n, 0)$ circle system with gcd(n,4) = 4, then the size of Ω is given by:* $q = \frac{(n-4)(n^2-2n+6)}{24}$. □

Now suppose Ω is a $(Z_n, 1)$ circle system with n even and such that gcd(n,3) = 1. Then (2) yields:

$\delta(x) = \tau(1-x) - \rho(1-2x) + \varepsilon$. Since $4x = 1$ has no solutions in Z_n, we have $\varepsilon = 1$ for all x. The expression 1-2x is odd, so by (4) we have $\rho(1-2x) = n/2$ for all x. Consider: $\tau(x)$ is the size of a (Z_n, x) triple system T. Since gcd(n,3) = 1, we have that T is equivalent to $(Z_n, 0)$ and has size (1/6)(n-1)(n-2) by Theorem 2. Thus:

$\delta(x) = (1/6)(n-1)(n-2) - (n/2) + 1 = \frac{(n-2)(n-4)}{6}$ *for all x in* Z_n *and thus:*

$\sum_{x \in Z_n} \delta(x) = \frac{n(n-2)(n-4)}{6}$ and $q = \frac{n(n-2)(n-4)}{24}$. This is the same count obtained for a $(Z_n, 1)$ circle system in which n|3, so we conclude:

Theorem 105 *If Ω is a $(Z_n, 1)$ circle system and n is even, then the size of Ω is given by* $q = \frac{n(n-2)(n-4)}{24}$. *If gcd(n,3) = 1 then Ω is regular of degree* $\frac{(n-2)(n-4)}{6}$. □

Now suppose Ω is a $(Z_n, 2)$ circle system with gcd(n,4) = 4 and gcd(n,3) = 1. Then from (2) we have:

$\delta(x) = \tau(2-x) - \rho(2-2x) + \varepsilon$. Since $4x = 2$ has no solutions in Z_n, we have $\varepsilon = 1$ for all x. The expression 2 - 2x is even, so by (4) we have $\rho(2-2x) = (1/2)(n-2)$. Since cd(n,3) = 1, we obtain $\tau(2-x) = (1/6)(n-1)(n-2)$ for all x. Thus:

$\delta(x) = (1/6)(n-1)(n-2) - (1/2)(n-2) + 1 = \frac{n^2-6n+14}{6}$ *for all x. Then from (1) we obtain:*

$\sum_{x \in Z_n} \delta(x) = \frac{n(n^2-6n+14)}{6}$, *and so* $q = \frac{n(n^2-6n+14)}{24}$.

Conclusion: If Ω is a $(Z_n, 2)$ circle system with gcd(n,4) = 4 and gcd(n,3) = 1, then Ω is regular of degree $\frac{n^2-6n+14}{6}$ and the size of Ω is given by $q = \frac{n(n^2-6n+14)}{24}$.

Finally suppose Ω is a $(Z_n, 2)$ circle system with gcd(n,4) = 4 and such that 3|n. Then from (2) we have:

$\delta(x) = \tau(2-x) - \rho(2-2x) + \varepsilon$. Again, since $4x = 2$ has no solutions in Z_n, we have $\varepsilon = 1$ for all x, and since 2 - 2x is even we have $\rho(2-2x) = (1/2)(n-2)$ for all x. From Theorem 3 we have $\tau(2\text{-}x) = (1/6)(n^2 - 3n + 6)$ if 3|(2-x), and $\tau(2\text{-}x) = (1/6)(n^2 - 3n)$ otherwise. This yields $\delta(x) = (1/6)(n^2 - 3n + 6) - (1/2)(n-2) + 1 = \frac{n^2-6n+18}{6}$ if 3|(2-x), and $\delta(x) = (1/6)(n^2 - 3n) - (1/2)(n-2) + 1 = \frac{n^2-6n+12}{6}$ otherwise.

Since there are n/3 elements in Z_n which satisfy 3|(n(2-x) we obtain:

$\sum_{x \in Z_n} \delta(x) = (n/3)(\frac{n^2-6n+18}{6}) + (2n/3)(\frac{n^2-6n+12}{6}) = \frac{n(n^2-6n+14)}{6}$, and thus $q = \frac{n(n^2-6n+14)}{24}$. This is the same count as in the previous case so we conclude:

Theorem 106 *If Ω is a $(Z_n, 2)$ circle system with $gcd(n,4) = 4$, then the size of Ω is given by* $q = \frac{n(n^2-6n+14)}{24}$. □

10.2 Sizes of Circle Systems with Base $Z_2 \times Z_n$

Let Ω be $(Z_2 \times Z_n, (a,b))$ circle system, then for $(x,y) \in Z_2 \times Z_n$ equation (2) takes the form:

(6) $\delta(x,y) = \tau(a\text{-}x, b\text{-}y) - \rho(a\text{-}2x, b\text{-}2y) + \varepsilon$, *where* $\varepsilon = 0$ *if* $4(x,y) = (a,b)$ *and otherwise* $\varepsilon = 1$.

Since $2x = 0$ and $4x = 0$ this can be simplified to:

(7) $\delta(x,y) = \tau(a\text{-}x, b\text{-}y) - \rho(a, b\text{-}2y) + \varepsilon$, *where* $\varepsilon = 0$ *if* $4y = b$.

In what follows we can always assume n is even, since if n is odd then $Z_2 \times Z_n$ is isomorphic to Z_{2n} whose size has been determined above. We will make use of:

Theorem 107 *Let Ω be a circle system with base $Z_2 \times Z_n$ where $n = 2m$.*

(i) If $gcd(m,4) = 1$, then Ω is equivalent to either a $(Z_2 \times Z_n, (0,0))$ or a $(Z_2 \times Z_n, (0,1))$ circle system.

(ii) If $gcd(m,4) = 2$, then Ω is equivalent to either a $(Z_2 \times Z_n, (0,0))$, $(Z_2 \times Z_n, (0,1))$, $(Z_2 \times Z_n, (0,2))$, or a $(Z_2 \times Z_n, (1,0))$ circle system.

(iii) If $gcd(m,4) = 4$, then Ω is equivalent to either a $(Z_2 \times Z_n, (0,0))$, $(Z_2 \times Z_n, (0,1))$, $(Z_2 \times Z_n, (0,2))$, $(Z_2 \times Z_n, (1,0))$ or a $(Z_2 \times Z_n, (1,2))$ circle system. □

Theorem 108 *If Ω is a $(Z_2 \times Z_n, (1,y))$ circle system, then the size of Ω is given by* $q = 2\binom{n}{3} = \frac{n(n-1)(n-2)}{3}$ *for any $y \in Z_n$.*

Proof Circles in Ω have the form ((0,a),(0,b),(0,c),(1,d)) or ((1,a),(1,b),(1,c),(0,d)) . Clearly there is a one-one correspondence between circles of these two types. There are $\binom{n}{3}$ ways to select a, b, c and once these are selected the last point (1,d) is forced. Thus $q = 2\binom{n}{3}$ as stipulated. □

In evaluating (7) the following observation is useful:

Theorem 109 *Let Ω be a circle system with base $Z_2 \times Z_n$, (n even), then $\rho(0,y) = n - 2$ if y is even; $\rho(0,y) = n$ if y is odd, and $\rho(1,y) = n$ for all y in Z_n.* □

Concerning triple systems with base $Z_2 \times Z_n$, we have the following:

Theorem 110 *Let T be a triple system with base $Z_2 \times Z_n$, (n even), then:*

(i) If $gcd(n,3) = 1$ then T is equivalent to a $(Z_2 \times Z_n, (0,0))$ triple system and has size $\frac{(n-1)(2n-1)}{3}$.

(ii) If $3/n$ and $3/y$, then a $(Z_2 \times Z_n, (0,y))$ or a $(Z_2 \times Z_n, (1,y))$ triple system is equivalent to a $(Z_2 \times Z_n, (0,0))$ triple system and has size $\frac{2n^2-3n+3}{3}$. □

If $3|n$ and $gcd(3,y) = 1$, then a $(Z_2 \times Z_n, (0,y))$ or a $(Z_2 \times Z_n, (1,y))$ triple system is equivalent to a $(Z_2 \times Z_n, (0,1))$ triple system and has size $\frac{2n^2-3n}{3}$. From Theorem 110 we obtain the following results which are useful in evaluating (7):

Theorem 111 *Let Ω be a circle system with base $Z_2 \times Z_n$, (n even), then for all y in Z_n:*

(i) If $gcd(n,3) = 1$, then $\tau(0,y) = \tau(1,y) = \frac{(n-1)(2n-1)}{3}$

(ii) If $3/n$ and $3/y$ then $\tau(0,y) = \tau(1,y) = \frac{2n^2-3n+3}{3}$

(iii) If $3/n$ and $gcd(y,3) = 1$, then $\tau(0,y) = \tau(1,y) = \frac{2n^2-3n}{3}$ □

Let Ω be a $(Z_2 \times Z_n, (0,0))$ circle system with $n = 2m$ and m odd. Then (7) yields $\delta(x,y) = \tau(-x,-y) - \rho(0,-2y) + \varepsilon$, where $\varepsilon = 1$ if $4(x,y) = (0,0)$ and $\varepsilon = 0$ otherwise. Thus $\varepsilon = 0$ if $(x,y) \in \{ (0,0), (1,0), (0,m), (1,m) \}$, and since $\rho(0,-2y) = n-2$ by Theorem 109. So we have $\delta(x,y) = \tau(-x,-y) - (n-2)$ if $(x,y) \in \{ (0,0), (1,0), (0,m), (1,m) \}$, and $\delta(x,y) = \tau(-x,-y) - (n-2) + 1$ otherwise. Suppose $gcd(n,3) = 1$. Then $\tau(-x,-y) = \frac{(n-1)(2n-1)}{3}$ for all $(x,y) \in Z_2 \times Z_n$ by Theorem 110, thus $\delta(x,y) = \frac{(n-1)(2n-1)}{3} - (n-2) = \frac{2n^2-6n+7}{3}$ if $(x,y) \in \{ (0,0), (1,0), (0,m), (1,m) \}$, and $\delta(x,y) = \frac{(n-1)(2n-1)}{3} - (n-2) + 1 = \frac{2n^2-6n+10}{3}$ otherwise. Then:

$$\sum_{(x,y)\in Z_2\times Z_n} \delta(x,y) = 4(\frac{2n^2-6n+7}{3}) + (2n-4)(\frac{2n^2-6n+10}{3}) = \frac{4(n-1)(n^2-2n+3)}{3},$$

and so $q = \frac{(n-1)(n^2-2n+3)}{3}$. Now suppose $3|n$. Then $\tau(-x,-y) = \frac{2n^2-3n+3}{3}$ if $3|y$, and $\tau(-x,-y) = \frac{2n^2-3n}{3}$ otherwise by Theorem 111. Then we have $\delta(x,y) = \frac{2n^2-3n+3}{3} - (n-2) = \frac{2n^2-6n+9}{3}$ if $(x,y) \in \{ (0,0), (1,0), (0,m), (1,m) \}$, (note that $3|y$ in this case.); $\delta(x,y) = \frac{2n^2-3n+3}{3} - (n-2) + 1 = \frac{2n^2-6n+12}{3}$ if $3|y$ and $y \notin \{0,m\}$, and $\delta(x,y) = \frac{2n^2-3n}{3} - (n-2) + 1 = \frac{2n^2-6n+9}{3}$ if $gcd(3,y) = 1$.

Since there are $2n/3$ elements (x,y) in $Z_2 \times Z_n$ such that $3|y$, we obtain:

$\sum_{(x,y)\in Z_2\times Z_n}\delta(x,y) = 4(\frac{2n^2-6n+9}{3}) + (\frac{2n}{3}-4)(\frac{2n^2-6n+12}{3}) + (\frac{4n}{3})(\frac{2n^2-6n+9}{3}) = \frac{4(n-1)(n^2-2n+3)}{3}$,

and so $q = \frac{(n-1)(n^2-2n+3)}{3}$. This is the same circle count as in the case gcd(n,3) = 1, so we conclude:

Theorem 112 *If Ω is a ($Z_2 \times Z_n$, (0,0)) circle system with n = 2m and m odd, then the size of Ω is given by* $q = \frac{(n-1)(n^2-2n+3)}{3}$.

Now let Ω be a ($Z_2 \times Z_n$, (0,0)) circle system with n = 2m and m even (m = 2q). Then (7) yields δ(x,y) = τ(-x,-y) - ρ(0,-2y) + ε, where ε = 0 if 4(x,y) =(0,0) and ε = 1 otherwise. The equation 4(x,y) = (0,0) now has eight solutions:

S = {(0,0),(1,0),(0,q), (1,q), (0,2q), (1,2q), (0,3q), (1,3q)},

and again ρ(0,-2y) = n-2 for all y by Theorem 7. So we have δ(x,y) = τ(-x,-y) - (n-2) if (x,y) ∈ S and δ(x,y) = τ(-x,-y) - (n-2) + 1 otherwise. Suppose gcd(n,3) = 1, then τ(-x,-y) = (1/3)(n-1)(2n-1) for all (x,y) ∈ $Z_2 \times Z_n$ and we obtain δ(x,y) = (1/3)($2n^2$ - 6n + 7) if (x,y) ∈ S, and δ(x,y) = (1/3)($2n^2$ - 6n + 10) otherwise. Then:

$$\sum_{(x,y)\in Z_2\times Z_n}\delta(x,y) = 8(1/3)(2n^2 - 6n + 7) + (2n-8)(1/3)(2n^2 - 6n + 10) =$$

$$(4/3)(n-2)(n^2 - n + 3),$$

with the consequence that the size q of Ω is given by q = (1/3)(n-2)(n^2 - n + 3). If 3|n, then we have from Theorem 110 that τ(-x,-y)= (1/3)($2n^2$ - 3n + 3) if 3|y, and τ(-x,-y)= (1/3)(n^2 - 3n) otherwise. Now we obtain δ(x,y) = (1/3) ($2n^2$ - 6n + 12) if 3|y and (x,y) ∉ S; δ(x,y) = (1/3) ($2n^2$ - 6n + 9) if (x,y) ∈ S. (note that 3|y in this case.); δ(x,y) = (1/3) ($2n^2$ - 6n + 9) if gcd(3,y) = 1. We then obtain:

$$\sum_{(x,y)\in Z_2\times Z_n}\delta(x,y) = \left(8+\left(\frac{4n}{3}\right)\right)\left(\frac{1}{3}\right)M + \left(\left(\frac{2n}{3}\right)-8\right)(M+1)$$

Where M = (1/3)($2n^2$ - 6n + 9) . This simplifies to:

$\sum_{(x,y)\in Z_2\times Z_n}\delta(x,y)$ = (4/3)(n-2)(n^2 - n + 3), yielding q = (1/3)(n-2)(n^2 - n + 3), the same result as for the case gcd(n,3) = 1. We conclude:

Theorem 113 *If Ω is a ($Z_2 \times Z_n$, (0,0)) circle system with n = 2m and m even, then the size q of Ω is given by q = (1/3)(n-2)(n^2 – n + 3).*

Let Ω be a ($Z_2 \times Z_n$, (0,1)) circle system with n even, (n = 2m). Suppose gcd(n,3) = 1. Then (7) yields δ(x,y) = τ(-x, 1-y) - ρ(0, 1-2y) + ε, where ε = 0 if 4(x,y) = (0,1) and ε = 1 otherwise. Since 4(x,y) = (0,1) can be satisfied only if 4y = 1, and 4y = 1 has no solutions in Z_n , we must have ε = 1 for all (x,y). Also, since 1-2y is odd, we have ρ(0, 1-2y) = n for all y by

Theorem 109. Finally, since gcd(n,3) = 1 we have $\tau(-x, 1-y) = \frac{(n-1)(2n-1)}{3}$ for all (x,y) in $Z_2 \times Z_n$. Thus $\delta(x,y) = \frac{(n-1)(2n-1)}{3} - n + 1 = \frac{2(n-1)(n-2)}{3}$ for all (x,y) in $Z_2 \times Z_n$. Then

$$\sum_{(x,y)\in Z_2\times Z_n} \delta(x,y) = 2n\left(\frac{2(n-1)(n-2)}{3}\right) \text{ and } q = \frac{n(n-1)(n-2)}{3}.$$

Now suppose 3|n. Then by Theorem 111 we have $\tau(0,y) = \tau(1,y) = \frac{2n^2-3n+3}{3}$ if 3|y and $\tau(0,y) = \tau(1,y) = \frac{2n^2-3n}{3}$ if gcd(y,3) = 1. Then $\delta(x,y) = \frac{2n^2-3n+3}{3} - n + 1 = \frac{2n^2-6n+6}{3}$ if 3|(1-y), and $\delta(x,y) = \frac{2n^2-3n}{3} - n + 1 = \frac{2n^2-6n+3}{3}$ otherwise. There are 2n/3 elements y in Z_n such that 3|(1-y) and consequently:

$$\sum_{(x,y)\in Z_2\times Z_n} \delta(x,y) = (2n/3)\left(\frac{2n^2-6n+6}{3}\right) + (4n/3)\left(\frac{2n^2-6n+3}{3}\right) = \frac{4n(n-1)(n-2)}{3},$$

which yields $q = \frac{n(n-1)(n-2)}{3}$, the same count as for the case gcd(n,3) = 1.

Theorem 114 *If Ω is a $(Z_2 \times Z_n, (0,1))$ circle system with n even, then the size of Ω is given by: $q = \frac{n(n-1)(n-2)}{3}$. If gcd(n,3) = 1, then Ω is regular of degree $\frac{2(n-1)(n-2)}{3}$.* □

Next let Ω be a $(Z_2 \times Z_n, (1,0))$ circle system with n even. We know by Theorem 108 that the size of Ω is given by $q = \frac{n(n-1)(n-2)}{3}$ which is the same count as the $(Z_2 \times Z_n, (0,1))$ circle system just examined, but let us determine the vertex degrees. We have from (7): $\delta(x,y) = \tau(1-x, -y) - \rho(1-2x, -2y) + \varepsilon = \tau(1-x, -y) - \rho(1, -2y) + \varepsilon$, where $\varepsilon = 0$ if 4(x,y) = (1,0) and $\varepsilon = 1$ otherwise. Since 4x = 1 has no solutions in Z_2, we must have $\varepsilon = 1$ for all (x,y). From Theorem 109 we have $\rho(1, -2y) = n$ for all y. So we have $\delta(x,y) = \tau(1-x, -y) - n + 1$. Now if we suppose that gcd(n,3) = 1, then by Theorem 111 we have $\tau(1-x, -y) = \frac{(n-1)(2n-1)}{3}$ for all $(x,y) \in Z_2 \times Z_n$ and thus $\delta(x,y) = \frac{(n-1)(2n-1)}{3} - n + 1 = \frac{2(n-1)(n-2)}{3}$ for all $(x,y) \in Z_2 \times Z_n$. If 3|n then from Theorem 111 we obtain $\tau(1-x, -y) = \frac{2n^2-3n+3}{3}$ if 3|y, and $\tau(1-x, -y) = \frac{2n^2-3n}{3}$ if gcd(3,y) = 1. Thus $\delta(x,y) = \frac{2n^2-3n+3}{3} - n + 1 = \frac{2n^2-6n+6}{3}$ if 3|y, and $\delta(x,y) = \frac{2n^2-3n}{3} - n + 1 = \frac{2n^2-6n+3}{3}$ otherwise. So we see that a $(Z_2 \times Z_n, (1,0))$ circle system and a $(Z_2 \times Z_n, (0,1))$ circle system have the same size and the same degree structure. If n = 4 the resulting circle systems are not equivalent since they have distinct *pair-degree* sequences. (If Ω is a (G,g) circle system and x,y ∈ G, then the *pair-degree* δδ{x,y} is the number of circles in Ω which contain both x and y.) Now let Ω be a $(Z_2 \times Z_n, (0,2))$ circle system with n = 4m. Then (7) yields $\delta(x,y) = \tau((0,2) - (x,y)) - \rho((0,2) - (2x,2y)) + \varepsilon = \tau(-x, 2-y) - \rho(0, 2-2y) + \varepsilon$ where $\varepsilon = 0$ if 4(x,y) = (0,2) and $\varepsilon = 1$ otherwise. Since 2 - 2y is even, we have $\rho(0, 2-2y) = n - 2$ by Theorem 109. Also, 4(x,y) = (0,2) holds iff 4y = 2 mod n. Since 4|n this equation has no solutions in Z_n, and thus $\varepsilon = 1$ for all $(x,y) \in Z_2 \times Z_n$. So we have $\delta(x,y) =$

τ (-x, 2 - y) – (n-2) + 1 for all (x,y) $\in Z_2 \times Z_n$. Suppose gcd(n,3) = 1, then by Theorem 111, τ (-x, 2 - y) = $\frac{(n-1)(2n-1)}{3}$ for all x $\in Z_2$ and for all y $\in Z_n$, so we obtain δ(x,y) = $\frac{(n-1)(2n-1)}{3}$ – (n-2) + 1 = $\frac{2n^2-6n+10}{3}$ for all (x,y) $\in Z_2 \times Z_n$. Then:

$$\sum_{(x,y)\in Z_2\times Z_n} \delta(x,y) = 2n(\frac{2n^2-6n+10}{3}) = \frac{4n(n^2-3n+5)}{3}$$

and thus q = $\frac{n(n^2-3n+5)}{3}$. Now suppose 3|n. Then by Theorem 111, τ (-x, 2 - y) = $\frac{2n^2-3n+3}{3}$ if 3|(2- y), and τ (-x, 2 - y) = $\frac{2n^2-3n}{3}$ otherwise. Thus δ(x,y) = $\frac{2n^2-3n+3}{3}$ - (n-2) + 1 = $\frac{2n^2-6n+12}{3}$ if 3|(2-y) and δ(x,y) = $\frac{2n^2-3n}{3}$ - (n-2) + 1 = $\frac{2n^2-6n+9}{3}$ otherwise. There are n/3 elements in Z_n which satisfy 3|(2-y) and thus there are 2n/3 elements (x,y) in $Z_2 \times Z_n$ which satisfy 3|(2-y). Consequently:

$$\sum_{(x,y)\in Z_2\times Z_n} \delta(x,y) = (2n/3)(\frac{2n^2-6n+12}{3}) + (4n/3)(\frac{2n^2-6n+9}{3}) = \frac{4n(n^2-3n+5)}{3}$$

and thus q = $\frac{n(n^2-3n+5)}{3}$. This is the same count as for the case gcd(n,3) = 1, so we conclude:

Theorem 115 *If Ω is a ($Z_2 \times Z_n$, (0,2)) circle system with n = 4m , then the size of Ω is given by* $q = \frac{n(n^2-3n+5)}{3}$. □

Finally suppose Ω is a ($Z_2 \times Z_n$, (1,2)) circle system with n = 8m. From (7) we obtain δ(x,y) = τ((1,2) – (x,y)) - ρ((1,2) – 2(x,y)) + ε = τ(1-x, 2-y) - ρ(1, 2 – 2y) + ε, where ε = 0 if 4(x,y) = (1,2) and ε = 1 otherwise. Since 4x = 1 has no solutions in Z_2 , we have ε = 1 for all (x,y). By Theorem 109, ρ(1, 2 – 2y) = n for all y in Z_n , so we have δ(x,y) = τ(1-x, 2-y) - n + 1. Now suppose gcd(n,3) = 1, then by Theorem 109, we have τ(1-x, 2-y) = $\frac{(n-1)(2n-1)}{3}$ for all (x,y) $\in Z_2 \times Z_n$, and so δ(x,y) = $\frac{(n-1)(2n-1)}{3}$ - n + 1 = $\frac{2(n-1)(n-2)}{3}$ for all (x,y) $\in Z_2 \times Z_n$, and so $\sum_{(x,y)\in Z_2 \times Z_n} \delta(x,y)$ = (2n)($\frac{2(n-1)(n-2)}{3}$), and q = $\frac{n(n-1)(n-2)}{3}$ in agreement with Theorem 108. Now suppose 3|n. Then we have by Theorem 111: δ(x,y) = $\frac{2n^2-3n+3}{3}$ - n + 1 = $\frac{2n^2-6n+6}{3}$ if 3|(2-y), and δ(x,y) = $\frac{2n^2-3n}{3}$ - n + 1 = $\frac{2n^2-6n+3}{3}$ otherwise. There are 2n/3 elements (x,y) in $Z_2 \times Z_n$ such that 3|(2-y) and so:

$$\sum_{(x,y)\in Z_2 \times Z_n} \delta(x,y) = (2n/3)(\frac{2n^2-6n+6}{3}) + (4n/3)(\frac{2n^2-6n+3}{3}) = \frac{4n(n-1)(n-2)}{3},$$

and thus q = $\frac{n(n-1)(n-2)}{3}$. Again this is in agreement with Theorem 108. Also we note that the degree sequence is the same as for a ($Z_2 \times Z_n$, (0,1)) or ($Z_2 \times Z_n$, (1,0)) circle system.

10.3 Using pair-degree sequences to determine geometric nonequivalence

We had previously remarked that a $(Z_2 \times Z_4, (0,1))$ circle system and a $(Z_2 \times Z_4, (1,0))$ circle system are not equivalent since they have different pair-degree sequences. This suggests that pair-degrees can be used to prove geometric nonequivalence for the general case of a $(Z_2 \times Z_n, (0,1))$ circle system and a $(Z_2 \times Z_n, (1,0))$ circle system, $(n = 2m)$. The general formula for pair degrees in a (G,g) circle system is:

(10) $\delta\delta\{a,b\} = \rho(g - a - b) - 2 + \eta + \mu + \lambda$, where $\eta = 1$ if $3a + b = g$, and $\eta = 0$ otherwise; $\mu = 1$ if $a + 3b = g$ and $\mu = 0$ otherwise; $\lambda = 1$ if $2a + 2b = g$ and $\lambda = 0$ otherwise.

The explanation for this formula is as follows. The remaining two elements in a circle of Ω which contains a and b must sum to $g - a - b$. The expression $\rho(g - a - b)$ counts the number of pairs in G which have this sum, but we must exclude from this count the pair which employs a and the pair which employs b. This explains the -2 in the formula. It may be that $a+b+a+a = g$ in which case we must add one back to the count, i.e., $\eta = 1$. It also may be that $a+b+b+b = g$, in which case we must add one back to the count, ie., $\mu = 1$. Finally, it may be that $a+b+a+b = g$, in which case we must also add one back to the count , i.e., $\lambda = 1$.

Theorem 116 *Let Γ be a $(Z_2 \times Z_n, (1,0))$ circle system with $n = 2m$ and $m = 2q$, then:*

(i) If q is even there are $mn - 2n + 8$ pairs $\{a,b\}$ in $Z_2 \times Z_n$ with $\delta\delta\{a,b\} = n - 4$; $2(n-8)$ pairs $\{a,b\}$ with $\delta\delta\{a,b\} = n - 3$ and $n^2 - n + mn + 8$ pairs $\{a,b\}$ with $\delta\delta\{a,b\} = n - 2$.

(ii) If q is odd there are $mn - 2n + 4$ pairs $\{a,b\}$ in $Z_2 \times Z_n$ with $\delta\delta\{a,b\} = n-4$; $2(n-4)$ pairs $\{a,b\}$ with $\delta\delta\{a,b\} = n-3$ and $n^2 - n + mn + 4$ pairs $\{a,b\}$ with $\delta\delta\{a,b\} = n - 2$.

Proof: From formula (10) we have: $\delta\delta\{(0,x),(0,y)\} = \rho(1, -x-y) - 2 + \eta + \mu + \lambda$. Since $2(0,x) + 2(0,y) = (1,0)$ has no solutions for $x,y \in Z_n$, we have $\lambda = 0$. Also $3(0,x) + (0,y) = (1,0)$ and $(0,x) + 3(0,y) = (1,0)$ have no solutions $x,y \in Z_n$, and so we have $\eta = \mu = 0$ as well. By Theorem 109, $\rho(1, -x-y) = n$ for all $x,y \in Z_n$. We conclude: $\delta\delta\{(0,x),(0,y)\} = n-2$ for all $x,y \in Z_n$. Similarly: $\delta\delta\{(1,x),(1,y)\} = n-2$ for all $x,y \in Z_n$. Now consider pairs of the type $\{(0,x),(1,y)\}$. Formula (10) yields: $\delta\delta\{(0,x),(1,y)\} = \rho(0, -x-y) - 2 + \eta + \mu + \lambda$. The equation: $2(0,x) + 2(1,y) = (1,0)$ has no solutions for $x,y \in Z_n$, so have $\lambda = 0$. We obtain $\eta = 1$ if $3x+y = 0$, and $\mu = 1$ if $x+3y = 0$, and otherwise η and μ are 0. From Theorem 109 we have $\rho(0, -x-y) = n-2$ if $x+y$ is even, and $\rho(0, -x-y) = n$ if $x+y$ is odd. Putting these observations together, we obtain: $\delta\delta\{(0,x),(1,y)\} = n-4$ if $x+y$ is even and $3x+y \neq 0$, and $x+3y \neq 0$;

$\delta\delta\{(0,x),(1,y)\} =$ n-3 if exactly one of 3x+y, x+3y is 0, and $\delta\delta\{(0,x),(1,y)\} =$ n-2 if 3x+y = 0 = x+3y, or x+y is odd.

There are clearly $2\binom{n}{2} = n(n-1)$ pairs of the form {(0,x),(0,y)} or {(1,x),(1,y)} in $Z_2 \times Z_n$. Next we count the pairs of the form {(0,x),(1,y)} where 3x+y = x+3y = 0. Note that for these pairs, x+y must be even. Consider 3x+y = x+3y implies 2x = 2y. This last equation is satisfied only if y = x or y = x+m. If y = x, then we have 4x = 0 which has precisely the four solutions: {0, q, 2q, 3q}. If y = x+m then we have 4x + m = 0, i.e., 4x = m = 2q. If q is odd then this equation has no solutions, and if q is even, say q = 2t, then there are precisely the four solutions: {t, t+q, t+2q, t+3q}. We conclude: if q is odd then there are exactly 4 pairs of the form {(0,x),(1,y)} such that 3x+y = x+3y = 0, and if q is even then there are exactly 8 pairs of the form {(0,x),(1,y)} such that 3x+y = x+3y = 0.

There are nm ordered pairs (x,y) with $x,y \in Z_n$, and such that x+y is odd. So we obtain: if q is odd then there are mn + 4 pairs of the form {(0,x),(1,y)} such that 3x+y = x+3y = 0 or x+y is odd, and there are mn + 8 pairs of the form {(0,x),(1,y)} such that 3x+y = x+3y = 0 or x+y is odd.

Next we count the number of pairs of the form {(0,x),(1,y)} in $Z_2 \times Z_n$ such that exactly one of x+3y, 3x+y is 0. First suppose gcd(n,3) = 1, then for each $x \in Z_n$, there is exactly one value for y such that x+3y = 0, and there is exactly one value for y such that 3x+y = 0. There are thus n ordered pairs (x,y) which satisfy x+3y = 0, and n ordered pairs (x,y) such that 3x+y = 0. An ordered pair occurs in both lists iff x+3y = 3x+y = 0. We have just counted these above to obtain 4 pairs if q is odd and 8 if q is even. Suppose gcd(n,3) = 1. If q is odd then there are 2(n-4) pairs of the form {(0,x),(1,y)} such that exactly one of x+3y, 3x+y is 0, and if q is even there are 2(n-8) such pairs in $Z_2 \times Z_n$ (n = 4q).

Now suppose 3|n, then x+3y = 0 (mod n) has solutions iff x is a multiple of 3, and for each such value of x, there are exactly three corresponding values for y which satisfy this equation. This yields $\left(\frac{n}{3}\right)3 = n$ ordered pairs (x,y) which satisfy x+3y = 0. From this list we must remove those ordered pairs (x,y) for which x+3y = 0 = 3x+y. If q is odd, we have seen that this can happen iff 4x = 0, i.e., $x \in \{0,q,2q,3q\}$, and if q is even then there are an additional four ordered pairs which satisfy x+3y = 0 = 3x+y. So if q is odd there are n-4 ordered pairs (x,y) such that x+3y = 0 and $3x+y \neq 0$, and if q is even there are n-8 such ordered pairs. Similarly, if q is odd, there are n-4 ordered pairs (x,y) which satisfy 3x+y = 0 and $x+3y \neq 0$, and if q is even there are n-8 such ordered pairs. We conclude:

Suppose 3|n. If q is odd then there are 2(n-4) pairs of the form {(0,x),(1,y)} such that exactly one of x+3y, 3x+y is 0, and if q is even there are 2(n-8) such pairs in $Z_2 \times Z_n$ (n = 4q).

Note that this is precisely the same count of these pairs as in the case gcd(n,3) = 1. It remains only to count the number of pairs of the form {(0,x),(1,y)} such that x+y is even and $x+3y \neq 0$ and $3x+y \neq 0$. If q is even the count must be mn - 2n + 8, and if q is odd the

count must be nm - 2n + 4 in order to account for all $\binom{2n}{2} = 2n^2 - n$ two element subsets of $Z_2 \times Z_n$. Now by adding counts we obtain the results stated in the Theorem. □
□

We will now undertake the same analysis for a $(Z_2 \times Z_n, (0,1))$ circle system Ω where $n = 2m$ and $m = 2q$. We will apply formula (10) first to pairs of the form $\{(0,x),(1,y)\}$: $\delta\delta\{(0,x),(1,y)\} = \rho(1, 1\text{-}x\text{-}y) - 2 + \eta + \mu + \lambda$. By Theorem 109, $\rho(1, 1\text{-}x\text{-}y) = n$ for all $x,y \in Z_n$. The equation $2(0,x) + 2(1,y) = (0,1)$ has no solutions for $x,y \in Z_n$, and thus $\lambda = 0$. The equations $(0,x) + 3(1,y) = (0,1)$ and $3(0,x) + (1,y) = (0,1)$ also have no solutions for $x,y \in Z_n$, and thus $\eta = \mu = 0$ also. We conclude that $\delta\delta\{(0,x),(1,y)\} = n\text{-}2$ for all $x,y \in Z_n$.

Next we consider pairs of the form $\{(0,x),(0,y)\}$. From formula (10) we obtain: $\delta\delta\{(0,x),(0,y)\} = \rho(0, 1\text{-}x\text{-}y) - 2 + \eta + \mu + \lambda$. The equation $2(0,x) + 2(0,y) = (0,1)$ has no solutions for $x,y \in Z_n$, and thus $\lambda = 0$. The equation the first case we obtain $4x = 1$ which has no solution in Z_n, and in the second case we obtain $4x + m = 1$ which also has no solution since $m = 2q$ is even. Thus we cannot have that both $3x+y = 1$ and $x+3y = 1$. By Theorem 109, $\rho(0, 1\text{-}x\text{-}y) = n\text{-}2$ if x+y is odd, and $\rho(0, 1\text{-}x\text{-}y) = n$ if x+y is even. A similar analysis holds for pairs of type $\{(1,x),(1,y)$, so we obtain:

(I) $\delta\delta\{(0,x),(0,y)\} = \delta\delta\{(1,x),(1,y)\} = n\text{-}4$ *if* $3x+y \neq 1$ *and* $x+3y \neq 1$ *and* $x+y$ *is odd.*

(II) $\delta\delta\{(0,x),(0,y)\} = \delta\delta\{(1,x),(1,y)\} = n\text{-}3$ *if* $3x+y = 1$ *or* $x+3y = 1$, *(we cannot have* $3x+y = 1 = x+3y$*).*

(III) $\delta\delta\{(0,x),(0,y)\} = \delta\delta\{(1,x),(1,y)\} = n\text{-}2$ *if* $x+y$ *is even. (in this case* $3x+y \neq 1$ *and* $x+3y \neq 1$*)*

(IV) $\delta\delta\{(0,x),(1,y)\} = n\text{-}2$ *for all* $x,y \in Z_n$.

There are clearly n^2 pairs of type (IV) and n(m-1) pairs of type (III). Now suppose gcd(n,3) = 1. Then for each $x \in Z_n$ there exists a unique $y \in Z_n$ such that $x+3y = 1$, and there exists a unique $y \in Z_n$ such that $3x+y = 1$. If the ordered pair (x,y) satisfies $x+3y = 1$, then (y,x) satisfies $3x+y = 1$, so we obtain a total of n pairs {x,y} which satisfy one or the other of these equations. Taking into account pairs of the two types $\{(0,x),(0,y)\}$, $\{(1,x),(1,y)\}$ we obtain a total of 2n pairs of type (II). Now suppose $3|n$ and $x,y \in Z_n$ satisfy $x+3y = 1 \pmod n$. This can only happen if $x = 3k+1$ for some integer k. There are n/3 such values in Z_n, and for each such value there are exactly three corresponding values of y such that $x+3y = 1$. This yields a total of $(n/3)3 = n$ ordered pairs (x,y) which satisfy this equation. Since we never have $x+3y = 3x+y = 1$, we conclude that there are exactly n ordered pairs (x,y) such that $x+3y = 1$ and $3x+y \neq 1$. Similarly there are exactly n ordered pairs (x,y) such thast $3x+y = 1$ and $x+3y \neq 1$. So we again obtain a total of 2n pairs of type (II). This leaves a total of $2(m^2 - n)$ pairs of Type (I). We then obtain the following result.

Theorem 117 *Let Ω be a $(Z_2 \times Z_n, (0,1))$ circle system, with $n = 2m = 4q$. Then Ω contains $2(m^2 - n)$ pairs $\{a,b\}$ with $\delta\delta\{a,b\} = n\text{-}4$; Ω contains $2n$ pairs with $\delta\delta\{a,b\} = n\text{-}3$, and $n(n+m\text{-}1)$ pairs with $\delta\delta\{a,b\} = n\text{-}2$.* □

By comparing the results of Theorems 116, 117 we easily obtain:

Theorem 118 *If Ω is a $(Z_2 \times Z_n, (0,1))$ circle system, with $n = 2m = 4q$ and Γ is a $(Z_2 \square Z_n, (1,0))$ circle system with $n = 2m = 4q$, then Ω and Γ are not geometrically equivalent.* □

Analogous to formula (1), we have for a circle system Ω with size q:

(13) $\sum_{\{x,y\} \subseteq \Omega} \delta\delta\{x, y\} = 6q.$

The justification for this formula is that in summing the pair degrees we count each circle in Ω six times since each circle contains $\binom{4}{2} = 6$ pairs. Thus if Ω is a $(Z_2 \times Z_n, (0,1))$ circle system with $n=2m = 4q$ and $\gcd(n,3) = 1$, then from Theorem 117 and formula 13 we obtain:

$$q = (1/6)\{2(m^2 - n)(n\text{-}4) + 2n(n\text{-}3) + n(n+m\text{-}1)(n\text{-}2)\} =$$

$$(1/6)\{(mn\text{-}2n)(n\text{-}4) + 2n^2 - 6n + n^3 - 2n^2 + n^2m - 2nm - n^2 + 2n\} =$$

$$(1/6)\{mn^2 - 4mn - 2n^2 + 8n + 2n^2 - 6n + n^3 - 2n^2 + n^2m - 2nm - n^2 + 2n\} =$$

$$(1/6)\{2n^2m - 6mn + 4n - 3n^2 + n^3\} = (1/6)\{2n^3 - 6n^2 + 4n\} = (1/3)n(n\text{-}1)(n\text{-}2).$$

which is in agreement with our earlier determination. A similar computation can be used to verify the size of a $(Z_2 \times Z_n, (0,1))$ circle system with $n = 2m = 4q$ and $\gcd(n,3) = 1$. Let Λ denote a $(Z_2 \times Z_n, (1,2))$ circle system with $n = 2m = 4q = 8t$. We will use formula (10) to compute pair-degrees for Λ. Consider: $\delta\delta\{(0,x),(0,y)\} = \rho(1, 2\text{-}x\text{-}y) - 2 + \eta + \mu + \lambda$. None of the equations $3(0,x) + (0,y) = (1,2)$, $(0,x) + 3(0,y) = (1,2)$ or $2(0,x) + 2(0,y) = (1,2)$ have solutions for $x,y \in Z_n$, so we have $\eta = \mu = \lambda = 0$. Also, by Theorem 109, $\rho(1, 2\text{-}x\text{-}y) = n$ for all $x,y \in Z_n$, so we obtain: $\delta\delta\{(0,x),(0,y)\} = n\text{-}2$ for all $x,y \in Z_n$, and similarly $\delta\delta\{(1,x),(1,y)\} = n\text{-}2$ for all $x,y \in Z_n$. Also using formula (10) we obtain: $\delta\delta\{(0,x),(1,y)\} = \rho(0, 2\text{-}x\text{-}y) - 2 + \eta + \mu + \lambda$. By Theorem 109, $\rho(0, 2\text{-}x\text{-}y) = n\text{-}2$ if x+y is even, and $(0, 2\text{-}x\text{-}y) = n$ if x+y is odd. Consider: $3(0,x) + (1,y) = (1,2)$ iff $3x+y = 2$, and $(0,x)+3(1,y) = (1,2)$ iff $x+3y = 2$. If $x+3y = 3x+y = 2$, then we must have $2y = 2x$ which implies $y = x$ or $y = x+m$. If $y=x$ then we obtain $4x = 2 \pmod n$ which has no solutions since $8|n$. If $y = x+m$ then we obtain $4x+m = 2$, i.e., $4x+4t = 2 \pmod n$ which also has no solutions in Z_n. Thus we cannot have that both $x+3y$ and $3x+y$ equal 0. Also we note that $2(0,x) + 2(1,y) = (1,2)$ has no solutions in Z_n, and so $\lambda = 0$. We then obtain:

(I) $\delta\delta\{(0,x),(1,y)\} = n\text{-}2$ if $x+y$ is odd. (we cannot have $x+3y = 2$ or $3x+y = 2$.)
(II) $\delta\delta\{(0,x),(1,y)\} = n\text{-}4$ if $x+y$ is even and $x+3y \neq 2$ and $3x+y \neq 2$.

(III) $\delta\delta\{(0,x),(1,y)\} = n\text{-}3$ one of $x+3y$, $3x+y$ equals 2. (both cannot equal 2.)
(IV) $\delta\delta\{(0,x),(0,y)\} = n\text{-}2$ for all $x,y \in Z_n$.

There are clearly $2\binom{n}{2} = n(n\text{-}1)$ pairs of type IV, and nm pairs of type (I). We will now count the pairs of type (III). First suppose gcd(n,3) = 1, then for each $x \in Z_n$ there exist a unique $y \in Z_n$ such that $x+3y = 2$. We thus obtain a total of n ordered pairs (x,y) such that $x+3y = 2$. Similarly there are n such ordered pairs which satisfy $3x + y = 2$. So in this case we have a total of 2n pairs of type (III). Now suppose 3|n. Now for $x+3y = 2$ to have solutions, it must be that $x = 3k+2$ for some integer k. There are (n/3) such values for x in Z_n, and for each such x, there are 3 corresponding values of y so that $x+3y = 2$ is satisfied. Again we obtain $(n/3)3 = n$ ordered pairs (x,y) such that $x+3y = 2$, and similarly there are n ordered pairs such that $3x+y = 2$. We conclude that there are 2n pairs of type (III) whether gcd(n,3) = 1 or 3|n. This leaves a total of $n^2 \text{-}2n - nm$ pairs of type (II). Summing counts for types (I) and (IV) we obtain:

Theorem 119 *Let Λ be a $(Z_2 \times Z_n, (1,2))$ circle system with $n = 2m = 4q = 8t$. Then Λ contains $n^2 \text{-} 2n - nm$ pairs {a,b} with $\delta\delta\{a,b\} = n\text{-}4$; 2n pairs with $\delta\delta\{a,b\} = n\text{-}3$, and $n^2 - n + nm$ pairs with with $\delta\delta\{a,b\} = n\text{-}2$.*

By comparing Theorem 119 with Theorem 116, we see that a $(Z_2 \times Z_n, (1,2))$ circle system cannot be equivalent to a $(Z_2 \times Z_n, (1,0))$ circle system when 8|n. In Theorem 43 the $(Z_2 \times Z_{2m}, (1,2))$ circle system and the $(Z_2 \times Z_{2m}, (1,0))$ circle system are proved *algebraically nonequivalent* but no geometric distinction between the two systems was given. When a geometric distinction is found, e.g., differing pair degree sequences, then we say that the circle systems are *geometrically nonequivalent.*

Theorem 120 *A $(Z_2 \times Z_n, (1,2))$ circle system cannot be geometrically equivalent to a $(Z_2 \times Z_n, (1,0))$ circle system when 8/n.* □

But we observe that when 8|n , Theorem 117 and Theorem 119 give exactly the same result:

Theorem 121 *If 8/n then a $(Z_2 \times Z_n, (1,2))$ circle system Λ, and a $(Z_2 \times Z_n, (0,1))$ circle system Γ have the same size, the same degree sequence and the same pair-degree sequence.*
□

So although we have easily shown that Λ and Γ are not algebraically equivalent in Theorem 43, even the use of pair-degrees does not distinguish between these circle systems geometrically. By examining the various sizes for circle systems, we find, taking n = 4m, that The $(Z_{4m}, 1)$, $(Z_2 \times Z_{2m}, (0,1))$ and $(Z_2 \times Z_{2m}, (1,2))$ circle systems all have the same size given by (1/24)n(n-2)(n-4). We have shown that the latter two circle systems also have the same degree sequence and the same pair-degree sequence. To see how a $(Z_{4m}, 1)$, compares with these we will now determine its degree sequence. Let Ω be a $(Z_{2m}, 1)$ circle system with m even. If gcd(m,3) = 1 then we have found that Ω is regular with degree (2/3)(m-1)(m-2), and if $3|m$ then $\delta(x) = (2/3)(m^2 - 3m + 3)$ if $3|(1-x)$ and $\delta(x) = (1/3)(2m^2 - 6m + 3)$ otherwise. This is exactly the same degree sequence as a $(Z_2 \times Z_m, (0,1))$ circle system. We may yet hope to distinguish Ω from Λ and Γ by considering the pair-degree sequence for Ω: From (10) we have: $\delta\delta\{a,b\} = \rho(1-a-b) - 2 + \eta + \mu + \lambda$, where $\eta = 1$ if $3a+b = 1$ and $\eta = 0$ otherwise; $\mu = 1$ if $a+3b = 1$ and $\mu = 0$ otherwise; $\lambda = 1$ if $2a+2b = 1$ and $\lambda= 0$ otherwise. Since $n = 2m$ s even, the equation $2a+2b = 1$ has no solutions (mod n) and so we have $\lambda = 0$. If $a+3b = 3a+b = 1$, then we obtain $2b = 2a$ which yields $b = a$ or $b = a+m$. If $b=a$ then we have $4a = 1$ (mod n) which has no solutions, and if $b = a+m$ then we have $4a+m = 1$ (mod n) which also has no solutions since m is even. We conclude that $a+3b$ and $3a+b$ cannot both equal 1. From (4) we have $\rho(1-a-b) = m$ if $a+b$ is even, and $\rho(1-a-b) = m-1$ if $a+b$ is odd. So we obtain: $\delta\delta\{a,b\} = m-2$ if $a+b$ is even; $\delta\delta\{a,b\} = m-3$ if $a+b$ is odd and $a+3b \neq 1$ and $3a+b \neq 1$, and $\delta\delta\{a,b\} = m-2$ if $a+b$ is odd and one of $a+3b$, $3a+b$ equals 1. Suppose $\gcd(m,3) = 1$, then we find that there are exactly 2m pairs $\{a,b\}$ with $a,b \in Z_{2m}$ such that one of $a+3b$, $3a+b$ equals 1. Also there are $2\binom{m}{2} = m(m-1)$ pairs $\{a,b\}$ such that $a+b$ is even, (and so $a+3b \neq 1$ and $3a+b \neq 1$). From this we conclude:

If Ω is a $(Z_{2m},1)$ circle system with m even and $\gcd(m,3) = 1$, then Ω contains $m^2 + m$ pairs of vertices $\{a,b\}$ with $\delta\delta\{a,b\} = m-2$, and Ω contains $m^2 - 2m$ pairs $\{a,b\}$ with $\delta\delta\{a,b\} = m-3$.

We note that, with the exception of $m = 4$, a $(Z_{2m},1)$ circle system has a very different pair-degree sequence than a $(Z_2 \times Z_m, (0,1))$ circle system, at least when $\gcd(m,3) = 1$. We will consider the special case $m = 4$ later. For now we will determine the pair-degree sequence for Ω in case $3|m$. Now if $a+3b = 1$ then $a = 3k+1$ for some integer k. There are exactly $n/3$ such elements a in Z_n, and for each a there are three values for b in Z_n such that $a+3b = 1$ (mod n). This yields a total of $3(n/3) = n = 2m$ pairs $\{a,b\}$ such that $a+3b = 1$ or $3a+b = 1$. We thus obtain the same pair degree sequence as in the case $\gcd(m,3) = 1$. We conclude:

If Ω is a $(Z_{2m},1)$ circle system with m even, then Ω contains $m^2 + m$ pairs of vertices $\{a,b\}$ with $\delta\delta\{a,b\} = m-2$, and Ω contains $m^2 - 2m$ pairs $\{a,b\}$ with $\delta\delta\{a,b\} = m-3$.

Now regarding the case where $m = 4$, we have a $(Z_8, 1)$ circle system Ω and a $(Z_2 \times Z_4, (0,1))$ circle system Γ, each with the same size, the same degree sequence and the same pair-degree sequence, but these circle systems can be distinguished by considering the character of the pairs with pair-degree $m-3 = 1$. The pairs in Ω with pair-degree 1 are {{0,5}, {0,7}, {1,2}, {1,4}, {2,7}, {3,4}, {3,6}, {5,6}}. If these pairs are represented by a graph, we obtain a cycle of length 8. The pairs in Γ with pair-degree 1 are

{(0,0),(0,1)}, {(0,0),(0,3)}, {(1,3),(1,0)}, {(1,1),(1,0)}, {(0,2),(0,1)}, {(1,2),(1,1)}, (0,2),(0,3)}, {(1,2),(1,3)}

which form a graph consisting of two 4-cycles. From this we see that Ω and Γ are not geometrically equivalent. We can now state:

A (Z_{2m}, 1) circle system is not geometrically equivalent to a ($Z_2 \times Z_m$, (0,1)) or a ($Z_2 \times Z_m$, (1,2)) circle system for any even m.

It remains to consider the relationship between a ($Z_2 \times Z_n$ (0,1)) circle system Ω and a ($Z_2 \times Z_n$ (1,2)) circle system Λ when n is a multiple of 8. We have shown that Ω, Λ have the same size; the same degree sequence, and the same pair-degree sequence, but they are not algebraically equivalent. To determine geometric nonequivalence, we must look more deeply into their structure. First we consider the algebra (Ω, δ) where $\delta(a,b,c) = (0,1) - a - b - c$. The subset S of Ω defined by $S = \{(0,x): x \in Z_n\}$, and the subset defined by $T = \{(1,x): x \in Z_n\}$, are subalgebras of (Ω, δ) which partition Ω into two parts with the same cardinality. Moreover, S,T are clearly equivalent to (Z_n, 1) circle systems. Also S is a maximal subalgebra of (Ω, δ), for if A is a subalgebra of (Ω, δ) which properly contains S, then A must contain an element of the form (1,y). Let (1,x) be an arbitrary element of T, then $\delta((1,y),(0,-y),(0,1-x)) = (0,1) - (1,1-x) = (1,x)$ must lie in A, and it follows that $A = \Omega$. Similarly T is a maximal subalgebra. Suppose that B,C are maximal subalgebras of (Ω, δ) which partition Ω. It is easily seen that $S = \langle(0,0)\rangle$, i.e., the element (0,0) generates S. So if (0,0) lies in B, then we must have $B = S$ and $C = T$, and if (0,0) lies in C then we must have $C = S$ and $B = T$. Thus {S,T} is the unique partition of Ω into two maximal subalgebras.

There is a similar partition of Λ. Let $U = \{(0,x), (1,y) : x,y$ are even elements in $Z_n\}$, and let $W = \{(0,x), (1,y)$: x,y are odd elements in $Z_n\}$, then U,W are subalgebras of (Λ,δ) where δ is now defined by $\delta(a,b,c) = (1,2) - a - b - c$. The mapping $\phi : U \rightarrow Z_2 \times Z_m$ defined by $\phi(0,2s) = (0,s)$, and $\phi(1,2t) = (1,t)$ turns U into a ($Z_2 \times Z_m$,(1,1)) circle system which is, in turn, equivalent to a ($Z_2 \times Z_m$,(0,1)) circle system by Theorem B. We will now show that U is a maximal subalgebra of (Λ,δ). Suppose A is a subalgebra of (Λ,δ), and A properly contains U. Then A must contain either a point of the form (0,x) with x an odd element of Z_n, or a point (1,y) with y an odd element of Z_n. If $(0,x) \in A$ with x odd, then $\delta((0,x),(0,1-x), (0,1-y)) = (1,y)$ lies in A, for any odd y. Conversely, if (1,y) lies in A with y odd then $\delta((1,y),(0,1-x), (0,1-y)) = (0,x)$ lies in A for any odd x. As a consequence, A contains all the points in W and thus $A = \Lambda$. It follows that U, and similarly W are maximal subalgebras of (Λ,δ) which partition Λ into two parts with the same cardinality. If Ω and Λ are geometrically equivalent, then {U,W} must be the unique partition of Λ into two maximal subalgebras, and it must be that the ($Z_2 \times Z_m$,(0,1)) circle system obtained from U is equivalent to a (Z_n, 1) circle system, $(n = 2m)$. Since we are supposing that m is even, this is a direct contradiction to our previous result. So we may conclude:

If n is a multiple of 8, then a ($Z_2 \times Z_n$, (0,1)) circle system and a ($Z_2 \times Z_n$, (1,2)) circle system are not geometrically equivalent.

Chapter 11

One and Two Generator Subalgebras

11.1 One Generator Subalgebra of (γ,δ)

In Theorem 78 it was shown that every finite subalgebra of (γ,δ) is equivalent to a circle system with base Z_n or base $Z_2 \times Z_n$. In this Chapter we show that certain infinite subalgebras of (γ,δ) are also equivalent to circle systems. In this Section we consider a one-generator infinite subalgebra and show that it forms a (Z,1) circle system.

Theorem 122 *Let γ be an irreducible cubic curve which inverts onto itself via a circle with center X and let q be a point on γ. Label the points of* $\langle q \rangle_\delta$ *with integers as follows:* $q = 0$; $\varphi(q)=1$; $\delta(q, \varphi(q), \varphi(q)) = -1$; *If integer labels k-1, -k+1, (k>0), have been assigned to points on γ, then let* $\delta(q,q,-k+1) = k$ *and* $\delta(\varphi(q), \varphi(q), k-1) = -k$. *Let S denote the set of points so labeled and assume no point in S receives more than one label. Then S constitutes the full set of points in* $\langle q \rangle_\delta$ *and with this labeling* $\langle q \rangle_\delta$ *is a (Z,1) circle system.*

Proof We will first show that S is a perfect Z-gon. Define $X_0 = q*q$. We will show that X_0 is a perspective point by proving $k* -k = X_0$ for each integer k. An inductive argument will be used so we start by showing $1*-1 = X_0$. Consider:

$$1*-1 = \varphi(q) * \delta(q, \varphi(q), \varphi(q)) = \{X*((X*q)*(q*q))\} * \{X*((X* \varphi(q))*(q* \varphi(q))\} =$$

$$(X*X)* [\{(X*q)*(X* \varphi(q))\} * \{(q*q)*(q* \varphi(q))\}] =$$

$$(X*X)* [\{(X*X)*(q* \varphi(q))\} * \{(q*q)*(q* \varphi(q))\}] =$$

$$(X*X)* [\{(X*X)*(q* q)\} * \{(q* \varphi(q))*(q* \varphi(q))\}] =$$

$$(X*X)* [\{(X*X)*(q* q)\} * \{(q* q))*(\varphi(q)* \varphi(q))\}] =$$

$$(X*X)* [\{(X*X)*(\varphi(q)* \varphi(q))\} * \{(q* q))*(q*q)\}] =$$

$$(X*X)* [\{(X* \varphi(q))*(X* \varphi(q))\} * \{(q* q))*(q*q)\}] =$$

$$(X*X)* [\{(X* q)*(q*q))*(X*q)* (q*q))\} * \{(q* q))*(q*q)\}] =$$

$$(X*X)* [\{(X* q)*(X*q))*(q*q)* (q*q))\} * \{(q* q))*(q*q)\}] =$$

$$(X^*X)^*\{(X^*X)^*(q^*q)\} = q^*q = X_0$$

Now assume inductively that $s^* -s = X_0$ for s = k-1. To complete the induction we show that

$k^*-k = X_0$. Consider:

$$k^* -k = \delta(q,q,-k+1) * \delta(\varphi(q), \varphi(q), k-1) =$$

$$[X^* \{(X^*q)^*(q^* -k+1)\}] *[X^*\{(X^* \varphi(q))^*(\varphi(q)^* k-1)\}] =$$

$$(X^*X) * [\{(X^*q)^*(X^* \varphi(q))\} *\{(q^* -k+1)^*(\varphi(q)^* k-1)\}] =$$

$$(X^*X) * [\{(X^*q)^*(X^* \varphi(q))\} *\{(q^* \varphi(q))^*((-k+1)^* (k-1))\}] =$$

$$(X^*X) * [\{(X^*q)^*(X^* \varphi(q))\} *\{(q^* \varphi(q))^*X_0\}] =$$

$$(X^*X) * [\{(X^*q)^*(X^* \varphi(q))\} *\{(q^* \varphi(q))^*(q^*q)\}] = q^*q = X_0.$$

The last line is the same as the second line in the demonstration that $1^* -1 = X_0$. We have shown that the point $X_0 = q^*q$ is a perspective point for the points in S. We define $X_k = 0^*k$ for each integer k. To show each of these is a perspective point we begin by showing that X_1 and X_{-1} are perspective points. To show that $s * (-s+1) = X_1$, we show first that $-s+1 = \delta(q, \varphi(q), s-1)$. Consider:

$$\delta(q, \varphi(q), s-1) = -s+1 \text{ iff } \{X^*\{(X^* \varphi(q))^*(q^* (s-1))\} = -s+1 \text{ iff}$$

$$\{(X^* \varphi(q))^*(q^* (s-1))\} = X^*(-s+1) \text{ iff}$$

$$X^* \varphi(q)) = (q^*(s-1))^*(X^*(-s+1)) = (X^*q)^*\{(s-1) *(-s+1)\} \text{ iff}$$

$$X^* \varphi(q) = (X^*q)^*(q^*q).$$

The last line is certainly true so we have shown $-s+1 = \delta(q, \varphi(q), s-1)$. Then:

$$s^*(-s+1) = \delta(q,q,-s+1) * \delta(q, \varphi(q), s-1) =$$

$$[X^*\{(X^*q)^*(q^*(-s+1))\}] * [X^*\{(X^* \varphi(q))^*(q^*(s-1))\}] =$$

$$(X^*X)^* [\{(X^*q)^*(X^* \varphi(q))\} * \{(q^*q)^*((-s+1)^*(s-1))\}] =$$

$$(X^*X)^* [\{(X^*q)^*(X^* \varphi(q))\} * \{(q^*q)^*(q^*q)\}] =$$

$$(X^*X)^* [\{(X^*q)^*\{(X^* q)^*(q^*q)\}\} * \{(q^*q)^*(q^*q)\}] =$$

$$(X^*X) * [(q^*q)^* \{(q^*q)^*(q^*q)\}] = (X^*X)^*(q^*q) =$$

$$\{X^*(X^*q)\} * \{X^*((X^*q)^*(q^*q))\} = q * \varphi(q) = 0^*1 = X_1.$$

We have proved that X_1 is a perspective point of S. To show that X_{-1} is a perspective point we can use the following argument. Suppose X_k, (k> 0), has been proved to be a perspective point, then -s * (s-k) = (X_0 * s)*(X_0 * (k-s)). Here we have used the recently proven fact that X_0 is a perspective point. Then -s * (s-k) = (X_0*X_0) * (s*(k-s)) = (X_0*X_0) *X_k = (X_0*X_0) *(0*k) = (X_0 *0)*(X_0 *k) = (X_0 *0)* -k = (X_0 *0)* (X_{-k}*0) = (0*0)*(X_0*X_{-k}) = X_{-k}, and thus X_{-k} is also a perspective point. Another useful fact is that δ(q, φ(q), s-k) = k-s which we now prove. Consider:

$$\delta(q, \varphi(q), s\text{-}k) = k\text{-}s \textit{ iff } X*\{(X*\varphi(q))*(q*(s\text{-}k))\} = k\text{-}s \textit{ iff}$$

$$\{(X*\varphi(q))*(q*(s\text{-}k))\} = X*(k\text{-}s) \textit{ iff}$$

$$X*\varphi(q) = (q*(s\text{-}k))*(X*(k\text{-}s)) = (X*q)*\{(s\text{-}k)*(k\text{-}s)\} = (X*q)*X_0 \textit{ iff}$$

$$X*\varphi(q) = (X*q)*(q*q) \text{ which is certainly true.}$$

Now assume inductively that X_{k-1} and thus X_{1-k} have been proved to be perspective points for S and consider:

$$s*(k\text{-}s) = \delta(q,q, 1\text{-}s) * \delta(q, \varphi(q), s\text{-}k) =$$

$$[X*\{(X*q)*(q*(1\text{-}s))\}] * [X*\{(X*\varphi(q))*(q*(s\text{-}k))\}] =$$

$$(X*X) * [\{(X*q)*(X*\varphi(q))\} * \{(q*q)*((1\text{-}s)*(s\text{-}k))\}] =$$

$$(X*X) * [\{(X*q)*(X*\varphi(q))\} * \{(q*q)*X_{1\text{-}k}\}] =$$

$$(X*X) * [\{(X*q)*((X*q)*(q*q))\} * \{(q*q)*X_{1\text{-}k}\}] =$$

$$(X*X) * [(q*q) * \{(q*q)*X_{1\text{-}k}\}] = (X*X)*X_{1\text{-}k} = (X*X)* (\text{-}k * 1) =$$

$$(X*X)* \{(X_0*k)*\varphi(q)\} = (X*X)*\{((q*q)*(q*X_k)) * \varphi(q)\} =$$

$$[X*((q*q)*(q*X_k))] * (X*\varphi(q)) = [X*((q*q)*(q*X_k))] *[(X*q)*(q*q)] = q*(q*X_k) = X_k.$$

It follows by induction that the points of S form a perfect Z-gon. By Theorem 34, S is a (Z,1) circle system and consequently a subalgebra of (γ,δ). Since q ∈ S, we have $\langle q \rangle_\delta \subseteq S$ and consequently $S = \langle q \rangle_\delta$. □

In Figure 80 we illustrate, on a Type V self-inversive cubic curve, some of the points and circles of the (Z,1) circle system generated by the point q.

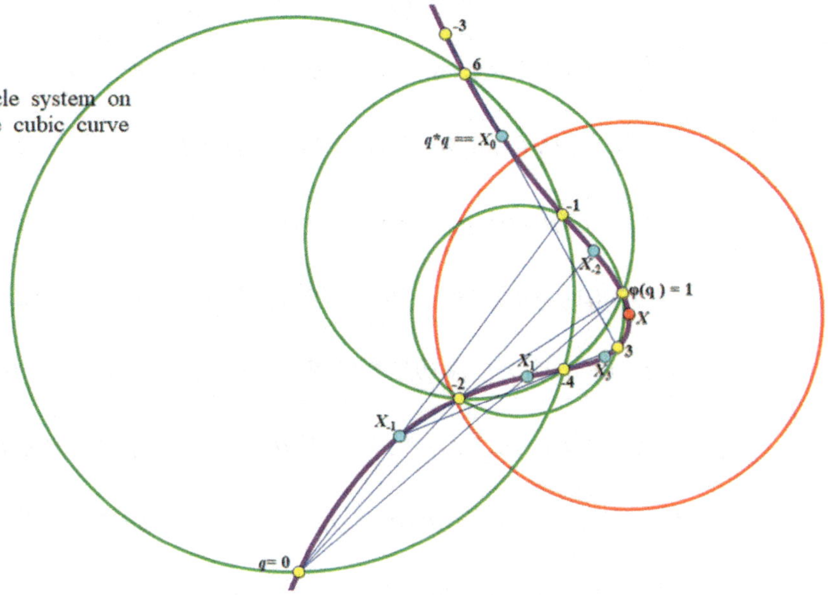

Figure 80 : (Z,1) circle system on Type V self-inversive cubic curve

11.2 Two Generator Subalgebra of (γ,δ)

In this Section we show that a two-generator infinite subalgebra of (γ,δ) is equivalent to a (Z×Z. (0,1)) circle system.

Theorem 123 *Let γ be an irreducible cubic curve which inverts onto itself via a circle with center X and let p, q be unrelated points on γ. Assign labels from Z×Z to points in* $\langle p,q\rangle_\delta$ *in accordance with the following scheme: q = (0,0); p = (1,0); φ(q) = (0,1); for k>0 let δ(q,q,(0,1-k) = (0,k) and δ(φ(q), φ(q), (0, k-1) = (0,-k); for h>0 let δ(φ(q), δ(p,q,φ(q)), (1-h,0)) = (h,0) and δ(p,φ(q),(h-1,0)) = (-h,0); for all integers h, k let δ((-h,0),(0,-k), φ(q)) = (h,k). Let S denote the set of points so labeled and assume no point in S receives more than one label. Then S constitutes the full set of points in* $\langle p,q\rangle_\delta$ *and with this labeling* $\langle p,q\rangle_\delta$ *is a (Z×Z, (0,1)) circle system.*

Proof We will follow the same proof procedure used in Theorem 122: first we show that S is a perfect Z×Z - gon and then use Theorem 34 to deduce that S is a circle system. We define $X_{0,0}$ = q*q and proceed to show step by step that $X_{0,0}$ is a perspective point for the set of points S. Consider:

$$(1,0)*(-1,0) = p*\,\delta(p,q,\varphi(q)) = p*\,[X*\{(X*\varphi(q))*(p*q)\}] =$$

$$p*\,[X*\,\{((X*q)*(q*q))\ *\ (p*q)\}] = p*\,[X*\,\{X\ *\ (p*(q*q))\}] =$$

$$p*\{p*(q*q)\} = q*q = X_{0,0}\,.$$

Now assume inductively that (s,0)*(-s,0) = $X_{0,0}$ for 0<s<k and consider:

$$(k,0)*(-k,0) = \delta(\varphi(q),\ \delta(p,q,\varphi(q)),\ (1-k,0)) * \delta(p,\varphi(q),\ (k-1,0)) =$$

$$\delta(\delta(q,q,q),\ \delta(p,q,\varphi(q)),\ (1-k,0)) * \delta(p,\varphi(q),\ (k-1,0)) =$$

$$\delta(\delta(p,q,q),\ \delta(q,q,\varphi(q)),\ (1-k,0)) * \delta(p,\varphi(q),\ (k-1,0)) =$$

$$\delta(\delta(p,q,q),\ q,\ (1-k,0)) * \delta(p,\varphi(q),\ (k-1,0)) =$$

$$[X*\{(X*\delta(p,q,q))*(q*(1-k,0)\}] * [X*\{(X*\varphi(q))*(p*(k-1,0))\}] =$$

$$[X*\{((X*p)*(q*q))*(q*(1-k,0)\}] * [X*\{((X*q)*(q*q)) * (p*(k-1,0))\}] =$$

$$[X*\{((X*p)*(1-k,0)) * q\}] * [X*\{((X*q)*(q*q)) * (p*(k-1,0))\}] =$$

$$(X*X)* [\{((X*p)*(1-k,0))*(p*(k-1,0))\} * \{q * ((X*q)*(q*q))\}] =$$

$$(X*X)* [\{((X*\{(1-k,0))*(k-1,0)\} * \{q * ((X*q)*(q*q))\}] =$$

$$(X*X)* [\{X*X_{0,0}\} * \{q * ((X*q)*(q*q))\}] =$$

$$(X*X)* [\{X*(q*q)\} * \{q * ((X*q)*(q*q))\}] =$$

$$(X*X)*\{(X*q)*(X*q)\} = q*q = X_{0,0}\,.$$

We have shown by induction that (k,0)*(-k,0) = $X_{0,0}$ for all integers k. The proof that (0,h)*(0,-h) = $X_{0,0}$ is essentially the same as the corresponding demonstration in Theorem 122. Now consider:

$$(h,k)*(-h,-k) = \delta((-h,0),(0,-k),\ \varphi(q)) * \delta((h,0),(0,k),\varphi(q)) =$$

$$[X * \{(X*\varphi(q))*((-h,0)*(0,-k))\}] * [X * \{(X*\varphi(q))*((h,0)*(0,k))\}] =$$

$$(X*X) * [\{(X*\varphi(q))*(X*\varphi(q))\} * \{((-h,0)*(h,0))*((0,-k)*(0,k))\}] =$$

$$(X*X) * [\{((X*q)*(q*q)) * ((X*q)*(q*q))\} * \{X_{0,0} * X_{0,0}\}] =$$

$$(X*X) * [[\{(X*q)*(X*q)\} * \{(q*q)*(q*q)\}] * \{(q*q)*(q*q)\}] =$$

$$(X*X)* \{(X*X)*(q*q)\} = q*q = X_{0,0}.$$

We have now completed the proof that $X_{0,0}$ is a perspective point of the set of points S. Next we define $X_{-1,0}$ = (0,0)*(-1,0) and prove that $X_{-1,0}$ is a perspective point of the set S. As a preliminary, we show that (1-s,0)*((s,0) = p*q. Consider:

$$(1\text{-}s,0)*((s,0) = \delta(\varphi(q),\ \delta(p,q,\varphi(q)),\ (s,0))\ *\ \delta(q,\ \varphi(q),\ (\text{-}s,0)) =$$

$$[X * \{(X*\delta(p,q,\varphi(q))\ *\ (\varphi(q)\ *(s,0))\}]\ *\ [X*\ \{(X*\varphi(q))*(q*(\text{-}s,0))\}] =$$

$$(X*X)\ *\ [\{(X*\delta(p,q,\varphi(q))\ *\ (X*\varphi(q)\}\ *\ \{(\varphi(q)\ *(s,0))*(q*(\text{-}s,0))\}] =$$

$$(X*X)\ *\ [\{(X*\delta(p,q,\varphi(q))\ *\ (X*\varphi(q)\}\ *\ \{(\varphi(q)\ *q)*((s,0)*(\text{-}s,0))\}] =$$

$$(X*X)*[\{(X*\varphi(q))*(p*q))*\ (X*\varphi(q)\}\ *\ \{(\varphi(q)\ *q)*X_{0,0}\}] =$$

$$(X*X)\ *\ [(p*q)\ *\ \{(\varphi(q)\ *q)*X_{0,0}\}] = (X*X)*\{\varphi(q)*(p*X_{0,0})\} =$$

$$(X*X)*[\{X*((X*q)*(q*q))\}*\{p*(q*q)\}] = (X*X)*\{(X*p)*(X*q)\} = p*q.$$

Now we are prepared to prove that $X_{-1,0}$ is a perspective point. Due to the repetitive nature of the computations, our presentation will be more terse. Consider:

$$(s\text{-}1,t)*(\text{-}s,\text{-}t) = \delta((1\text{-}s,0),(0,\text{-}t),\ \varphi(q))\ *\ \delta((s,0),(0,t),\varphi(q)) =$$

$$(X*X)*\ [\{(X*\varphi(q))*(X*\varphi(q))\}\ *\ \{((1\text{-}s,0)*(s,0))\ *\ ((0,\text{-}t)*(0,t))\}] =$$

$$(X*X)*\ [\{((X*X)*(q*q))\ *\ ((q*q)*(q*q))\}\ *\ \{(p*q)\ *\ (q*q)\}] =$$

$$(X*X)\ *\ [(X*X)*\{((q*q)*(q*q))*(p*q)] = q\ *\ \{p*(q*q)\} =$$

$$(0,0)*\ (p*X_{0,0}) = (0,0)*\ (\text{-}1,0) = X_{-1,0}.$$

In the course of the proof of Theorem 122 we argued that once X_0, X_1 are proved to be perspective points, then so is X_{-1}. More generally, once $X_{0,0}$, $X_{h,k}$ are proved to be perspective points in the present setting, then so is $X_{-h,-k}$. So we also have that $X_{1,0}$ is a perspective point of S. Now define $X_{h,0}$ = (0,0)*(h,0) and assume inductively that $X_{j,0}$ has been proved to be a perspective point of S for j = h-1 and j = 1-h. We will need (-s,0)*(s-h,0) = (0,0)*(-h,0) which we prove now. Consider:

$$(\text{-}s,0)*(s\text{-}h,0) = \delta(p,\varphi(q),(s\text{-}1,0))\ *\ \delta(q,\varphi(q),(h\text{-}s,0)) =$$

$$(X*X)*\ [\{(X*\varphi(q))*(X*\varphi(q))\}\ *\ \{(p*q)*((s\text{-}1,0)*(h\text{-}s,0))\}] =$$

$$(X*X)*\ [\{(X*\varphi(q))*(X*\varphi(q))\}\ *\ \{(p*q)*X_{h\text{-}1,0}\}] =$$

$$(X*X)\ *\ [\{((X*X)*(q*q))\ *\ (X_{0,0}*X_{0,0})\}\ *\ \{(p*q)*((\text{-}1,0)*(h,0))\}] =$$

$$(X*X)\ *\ [\{((X*X)*(q*q))\ *\ (X_{0,0}*X_{0,0})\}\ *\ \{(p*(\text{-}1,0)*(q*(h,0))\}] =$$

$$(X*X)\ *\ [\{((X*X)*(q*q))\ *\ (X_{0,0}*X_{0,0})\}\ *\ \{(X_{0,0}*(q*(h,0))\}] =$$

$$(X*X) * [\{((X*X)*(q*q)) * (X_{0,0}*X_{0,0})\} * \{((X_{0,0}*(X_{0,0}*X_{0,0}))*(q*(h,0))\}] =$$

$$(X*X) * [\{((X*X)*(q*q)) * (X_{0,0}*X_{0,0})\} * \{((q*(X_{0,0}*X_{0,0}))*(-h,0))\}] =$$

$$(X*X) * [\{((X*X)*(q*q)) * (-h,0))\} * q] =$$

$$(X*X) * [\{((X*X)*(q*q)) * (-h,0))\} * \{q*(q*q)\}] = q * (-h,0).$$

Now consider:

$$(s,t)*(h-s,-t) = \delta((-s,0),(0,-t),\varphi(q)) * \delta((s-h,0),(0,t),\varphi(q)) =$$

$$(X*X)* [\{(X*\varphi(q))*(X*\varphi(q))\} * \{((-s,0)*(s-h,0)) * ((0,-t)*(0,t))\}] =$$

$$(X*X)* [\{((X*X)*(q*q)) *(X_{0,0}*X_{0,0})\} * \{(q*(-h,0)) * X_{0,0}\}] =$$

$$(X*X)* [\{((X*X)*X_{0,0}) *(q*(-h,0))\} * X_{0,0}\}] =$$

$$(X*X)* [\{((X*X)*X_{0,0}) *(q*(-h,0))\} * \{X_{0,0}*(X_{0,0}*X_{0,0})\}] =$$

$$(X*X)* [\{((X*X) *\{(q*(-h,0))*(X_{0,0}*X_{0,0})\}\}] = (q*(-h,0))*(X_{0,0}*X_{0,0}) = q*(h,0) = X_{h,0}.$$

We have proved that $X_{h,0}$ is a perspective point for all integers. Next we define $X_{0,k} = q*(0,k)$ and prove that $X_{0,k}$ is a perspective point for S. To start the induction we show that $X_{0,1}$ is a perspective point. The following Claims will be useful. We leave the easy proofs of the first two to the reader.

Claim 1: $(0,b) = \delta(q,\varphi(q),(0,-b))$ for every integer b.

Claim 2: $(0,b)*\varphi(q) = (0, b+1)* q$ for every integer b.

Claim 3: $(0,t-1)*(0,-t) = q * \{X_{0,0} *\varphi(q)\}$.

Proof of Claim 3: We may suppose t > 0. Consider:

$$(0,t-1)*(0,-t) = \delta(q,\varphi(q),(0,1-t)) * \delta(\varphi(q),\varphi(q), (0,t-1)) =$$

$$(X*X)* [\{(X*\varphi(q))*(X*\varphi(q))\} * \{(q*\varphi(q))*X_{0,0}\}] =$$

$$(X*X)* [\{((X*q) * (X*q))* (X_{0,0}*X_{0,0})\} * \{(q*\varphi(q))*X_{0,0}\}] =$$

$$(X*X) * [\{((X*X)*X_{0,0}) * (q*\varphi(q))\} * X_{0,0}] =$$

$$(X*X) * [\{((X*X)*X_{0,0}) * (q*\varphi(q))\} * (X_{0,0}*(X_{0.0}*X_{0,0})] =$$

$$(X_{0.0}*X_{0,0})* (q*\varphi(q)) = q * \{X_{0,0} *\varphi(q)\}.$$

Now Consider:

$$(s,t)*(-s,\ 1\text{-}t) = \delta((-s,0),(0,-t),\varphi(q)) * \delta((s,0),(0,t\text{-}1),\varphi(q)) =$$

$$(X*X) * [\{((X*X)*X_{0,0})*(X_{0,0}*X_{0,0})\} * \{((-s,0)*(s,0))* ((0,-t)*(0,t\text{-}1))\}] =$$

$$(X*X) * [\{((X*X)*X_{0,0})*(X_{0,0}*X_{0,0})\} * \{X_{0,0}* (q* (X_{0,0} * \varphi(q))\}] =$$

$$(X*X)* [X_{0,0} * \{((X*X)*X_{0,0})*(q* (X_{0,0} * \varphi(q))\}] =$$

$$(X*X)* [X_{0,0} * \{((X*X)*q)) * \varphi(q)\}] =$$

$$(X*X)* [\{(q* X_{0,0})*q\} * \{((X*X)*q)) * \varphi(q)\}] =$$

$$(X*X)* [(X*X) * ((q*X_{0,0})*\varphi(q))] = q*\varphi(q) = X_{0,1}.$$

We have completed the proof that $X_{0,1}$ and thus $X_{0,-1}$ are perspective points for S. From Claim 2 we immediately obtain:

Claim 4: $(-s,\ k\text{-}t) = \delta(q,\ (s,0),(0,t\text{-}k\text{+}1))$

Proof of Claim 4: Consider:

$$(-s,\ k\text{-}t) = \delta(q,\ (s,0),(0,t\text{-}k\text{+}1)) \textit{ iff } \delta((s,0),(0,t\text{-}k),\varphi(q)) = \delta(q,\ (s,0),(0,t\text{-}k\text{+}1)) \textit{ iff}$$

$$(0,t\text{-}k)*\varphi(q) = (0,t\text{-}k\text{+}1)) * q$$

which follows from Claim 2. We assume inductively that $X_{0,k-1}$ = q*(0,k-1) and $X_{0,1-k}$ = q*(0,1-k) have been proved to be perspective points for S. Consider:

$$(s,t)*(-s,k\text{-}t) = \delta((-s,0),(0,-t),\varphi(q)) * \delta(q,\ (s,0),(0,t\text{-}k\text{+}1)) =$$

$$(X*X)* [\{(X*\varphi(q))*(X*q)\} * \{((-s,0)*(s,0))* ((0,-t)*(0,t\text{-}k\text{+}1))\}] =$$

$$(X*X)* [\{(X*\varphi(q))*(X*q)\} * \{X_{0,0} * X_{0,-k+1}\}]$$

$$(X*X)* [\{(X*\varphi(q))*(X*q)\}* \{X_{0,0}*(\varphi(q)*(0,-k))\}] =$$

$$(X*X)* [\{((X*q)*X_{0,0})*(X*q)\} * \{X_{0,0}*(\varphi(q)*(0,-k))\}] =$$

$$(X*X)*(\varphi(q)*(0,-k)) = (X*\varphi(q)) * (X*(0,-k) = ((X*q)*X_{0,0}) * (X*(0,-k)) =$$

$$q*(X_{0,0}*(0,-k)) = q*(0,k) = X_{0,k}.$$

We have now proved that $X_{0,k}$, and thus $X_{0,-k}$ are perspective points for the set S. Finally define $X_{h,k}$ = (h,0)*(0,k). We finish the proof of the Theorem by showing that $X_{h,k}$ is a perspective point. Consider:

$$(s,t)*(h-s,k-t) = \delta((-s,0),(0,-t),\varphi(q)) * \delta((s-h,0),(0,t-k),\varphi(q)) =$$

$$(X*X) *[\{((X*X)*(q*q))*(X_{0,0}*X_{0,0})\} * \{((-s,0)*(s-h,0)) * ((0,-t)*(0,t-k))\}] =$$

$$(X*X) *[\{((X*X)*(q*q))*(X_{0,0}*X_{0,0})\} * \{X_{-h,0} * X_{0,-k}\}] =$$

$$(X*X) *[\{((X*X)*(q*q))*(X_{0,0}*X_{0,0})\} * \{(q*(-h,0)) *((q*(0,-k))\}] =$$

$$(X*X) *[\{((X*X)*(q*q))*(X_{0,0}*X_{0,0})\} * \{(q*q) * ((-h,0)*(0,-k))\}] =$$

$$(X*X)*[(X*X) * \{(X_{0,0}*X_{0,0}) * ((-h,0)*(0,-k))\}] =$$

$$(X_{0,0} *(-h,0)) * (X_{0,0}*(0,-k)) = (h,0)*(0,k) = X_{h,k}.$$

This completes the proof that the points in S form a perfect Z×Z - gon, and so by Theorem 34, S is a (Z×Z, (0,1)) circle system. Then S is a subalgebra of (γ,δ) which contains both p and q and thus S contains $\langle p,q\rangle_\delta$. We then obtain S = $\langle p,q\rangle_\delta$. □

In Figure 81 we illustrate the (Z×Z, (0,1)) circle system generated by the points p,q on a self-inversive Type IV cubic curve.

Figure 81: (ZxZ,(0,1) circle system generated by p, q on a self-inversive Type IV cubic curve

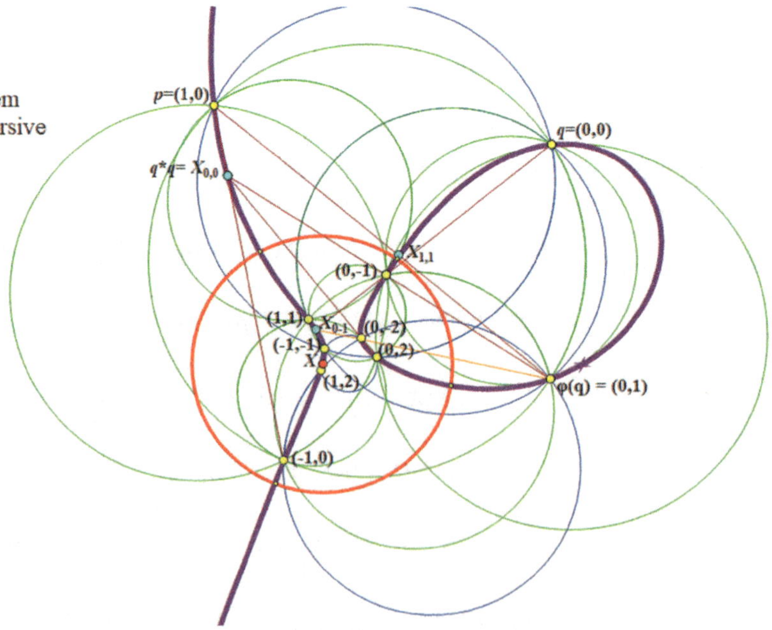

Appendix

Figure 82: Perfect closed pentagon with cubic envelope

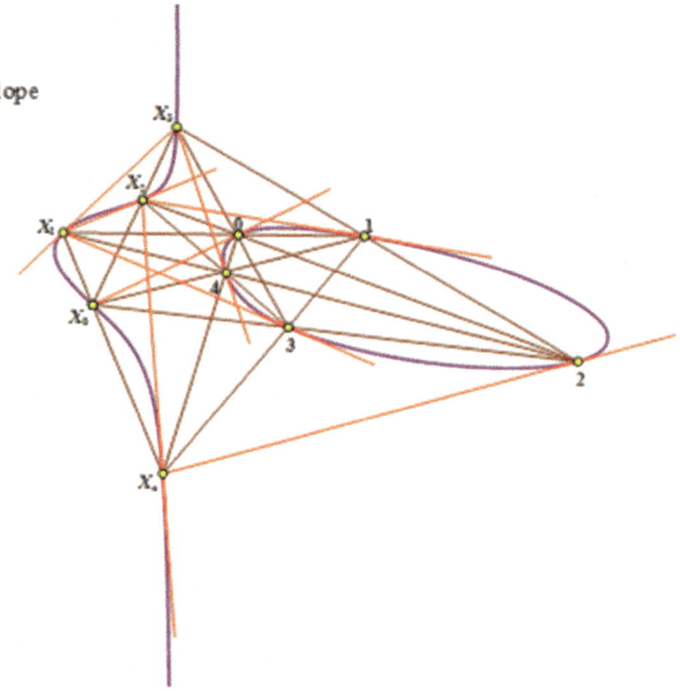

Figure 83: $(Z_{10},0)$ triple system with cubic envelope

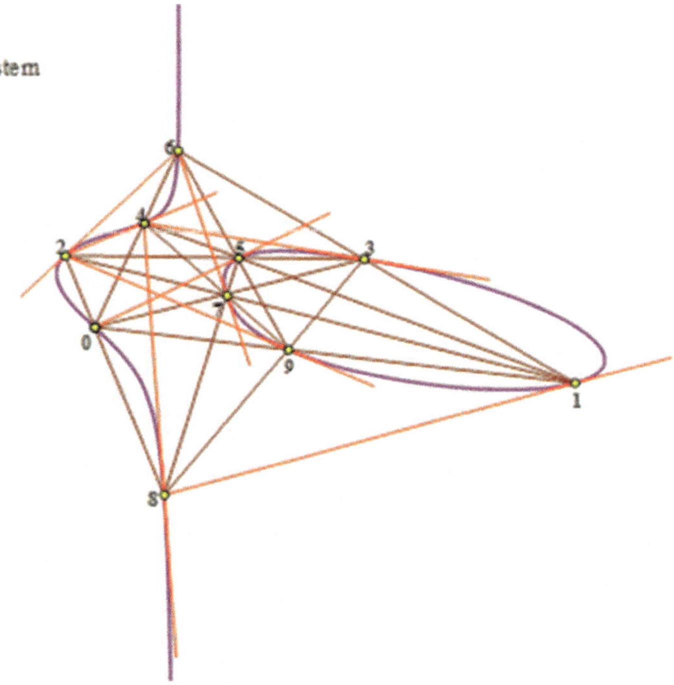

References

[1] Fletcher, R. : *Perfect Polygons and Geometric Triple Systems,* Springer Proceedings in Mathematics and Statistics (2016)

[2] Fletcher, R. : *Geometric Triple Systems with Base Z or Z_n,* Springer Proceedings in Mathematics and Statistics (2015)

[3] Bix, R. : *Conics and Cubics,* Springer, USA (2006)

[4] Fletcher, R. : *Group Circle Systems on Conics,* New Frontiers of Multidisciplinary Research in STEAM-H, Vol. 90, Springer International Publishing, Switzerland (2014)

[5] Hilton, H. : *Plane Algebraic Curves,* Oxford University Press, London (1920)

[6] Fletcher, R. : *Circle Chains,* J. Applied Global Research, Intellectbase International Consortium (2008)

[7] Fletcher, R. : *Self-inversive Cubic Curves,* Mathematical Sciences with Multidisciplinary Applications Vol. 157, Springer Internaional Publishing, Switzerland (2016)

[8] Coxeter, H.S.M and Greitzer, S.L., : *Geometry Revisited,* The Mathematical of America (1967)

[9] Coxeter, H.S.M, *Introduction to Geometry,* John Wiley & Sons Inc. (1969)

[10] Eves, Howard, *A Survey of Geometry, Vol. 1*, Allyn and Bacon, Inc., (1964)

[11] Brannan, David, Esplen, Matthew, Gray, Jeremy, *Geometry,* Cambridge University Press, new York, N.Y., (1999)

[12] Isaacs, Martin, *Algebra, a Graduate Course,* Brooks/Cole Publishing Co. (1994)

Index

Made in the USA
Monee, IL
02 November 2026

67385910R00090